Understanding Metaphors in the Life Sciences

Covering a range of metaphors from a diverse field of sciences, from cell and molecular biology to evolution, ecology, and biomedicine, *Understanding Metaphors in the Life Sciences* explores the positive and negative implications of the widespread use of metaphors in the biological and life sciences.

From genetic codes, programs, and blueprints, to cell factories, survival of the fittest, the tree of life, selfish genes, and ecological niches, to genome editing with CRISPR's molecular scissors, metaphors are ubiquitous and vital components of the modern life sciences. But how exactly do metaphors help scientists to understand the objects they study? How can they mislead both scientists and laypeople alike? And what should we all understand about the implications of science's reliance on metaphorical speech and thought for objective knowledge and adequate public policy informed by science?

This book will *literally* help you to better understand the *metaphorical* dimensions of science.

Andrew S. Reynolds is Professor of Philosophy at Cape Breton University, Canada. He is the author of *The Third Lens: Metaphor and the Creation of Modern Cell Biology* (University of Chicago Press, 2018) and *Peirce's Scientific Metaphysics: The Philosophy of Chance, Law and Evolution* (Vanderbilt University Press, 2002). He has a PhD in the philosophy of science from the University of Western Ontario.

The ***Understanding Life*** series is for anyone wanting an engaging and concise way into a key biological topic. Offering a multidisciplinary perspective, these accessible guides address common misconceptions and misunderstandings in a thoughtful way to help stimulate debate and encourage a more in-depth understanding. Written by leading thinkers in each field, these books are for anyone wanting an expert overview that will enable clearer thinking on each topic.

Series Editor: Kostas Kampourakis http://kampourakis.com

Published titles

Understanding Evolution	Kostas Kampourakis	9781108746083
Understanding Coronavirus	Raul Rabadan	9781108826716
Understanding Development	Alessandro Minelli	9781108799232
Understanding Evo-Devo	Wallace Arthur	9781108819466
Understanding Genes	Kostas Kampourakis	9781108812825
Understanding DNA Ancestry	Sheldon Krimsky	9781108816038
Understanding Intelligence	Ken Richardson	9781108940368
Understanding Metaphors in the Life Sciences	Andrew S. Reynolds	9781108940498

Forthcoming

Understanding Creationism	Glenn Branch	9781108927505
Understanding Species	John S. Wilkins	9781108987196
Understanding the Nature–Nurture Debate	Eric Turkheimer	9781108958165
Understanding How Science Explains the World	Kevin McCain	9781108995504
Understanding Cancer	Robin Hesketh	9781009005999
Understanding Forensic DNA	Suzanne Bell and John Butler	9781009044011
Understanding Race	Rob DeSalle and Ian Tattersall	9781009055581
Understanding Fertility	Gab Kovacs	9781009054164

Understanding Metaphors in the Life Sciences

ANDREW S. REYNOLDS
Cape Breton University

CAMBRIDGE
UNIVERSITY PRESS

University Printing House, Cambridge CB2 8BS, United Kingdom

One Liberty Plaza, 20th Floor, New York, NY 10006, USA

477 Williamstown Road, Port Melbourne, VIC 3207, Australia

314–321, 3rd Floor, Plot 3, Splendor Forum, Jasola District Centre,
New Delhi – 110025, India

103 Penang Road, #05–06/07, Visioncrest Commercial, Singapore 238467

Cambridge University Press is part of the University of Cambridge.

It furthers the University's mission by disseminating knowledge in the pursuit of education, learning, and research at the highest international levels of excellence.

www.cambridge.org
Information on this title: www.cambridge.org/9781108837286
DOI: 10.1017/9781108938778

First published 2022

Printed in the United Kingdom by TJ Books Limited, Padstow Cornwall

A catalogue record for this publication is available from the British Library.

Library of Congress Cataloging-in-Publication Data
Names: Reynolds, Andrew S., 1966– author.
Title: Understanding metaphors in the life sciences / Andrew S. Reynolds.
Description: Cambridge, United Kingdom ; New York, NY : Cambridge University Press, 2022. | Series: Understanding life | Includes bibliographical references and index.
Identifiers: LCCN 2021027054 (print) | LCCN 2021027055 (ebook) | ISBN 9781108837286 (hardback) | ISBN 9781108940498 (paperback) | ISBN 9781108938778 (ebook)
Subjects: LCSH: Communication in biology. | Metaphor. | Science – Language. | Life sciences – Philosophy. | BISAC: SCIENCE / Life Sciences / Genetics & Genomics
Classification: LCC QH303 .R48 2022 (print) | LCC QH303 (ebook) | DDC 570–dc23
LC record available at https://lccn.loc.gov/2021027054
LC ebook record available at https://lccn.loc.gov/2021027055

ISBN 978-1-108-83728-6 Hardback
ISBN 978-1-108-94049-8 Paperback

"What a timely book this is! It is precisely because biology has made such striking advances in recent years that its stock of metaphors is due for a clinical check-up. Reynolds offers a reliable and perceptive diagnosis of the framing narratives of the life sciences, sympathetically examining their strengths and weaknesses. This book should be an essential accompaniment to any study course in the biological sciences."

Philip Ball, science writer and author of *How to Grow a Human*

"In this beautifully written, highly accessible, and captivating work, Reynolds reveals the incredible extent to which scientific methods and descriptions in biology, the life sciences, and medicine are infused with metaphors. Interweaving the rich history and philosophy of the uses of these metaphors over time, their many implications for scientific reasoning, understanding, and the ethical and political dimensions of science itself are perceptively explored, with wonderful clarity and across an encyclopedic range of examples. Metaphors afford telling insight, opening doors to further inquiry and closing others. Is your genome software? Are enzymes molecular machines? Does nature select some traits over others, thereby constructing the tree of life? The fascinating world of metaphors in science comes to life on every page."

Anjan Chakravartty, University of Miami, USA

"I read Lakoff and Johnson's book *Metaphors We Live By* in the 1980s, and it was eye opening. Andrew Reynolds' book, which should be called *Metaphors Science Lives By*, is equally eye opening. Metaphors shape the way we live in the world. In science, they shape the way we understand the world. This can have huge implications for our lives, for better or for worse. How does this process of understanding work, especially in the life sciences? This book deals with the essential role of metaphors in this process. Written in an admirably clear style, Reynolds makes us aware of the power of metaphor, but also its dangers and pitfalls. It is an essential read for everybody interested in understanding how science and science communication work with and through metaphors. Importantly, it also dispels some common misunderstandings about the role of metaphors in science."

Brigitte Nerlich, University of Nottingham, UK

"*Understanding Metaphors in the Life Sciences* takes us from genes to cells, and up to the vast evolutionary tree of life, showing how science depends overwhelmingly on metaphor for understanding, for advance, for communication. A very important book."

Michael Ruse, Florida State University, USA

To my older siblings: Tina, Rhys, Anne, and Peter; for perpetually spoiling your baby brother while also molding me into a reasonably responsible and productive human being.

Contents

Preface

Even if you are not a working scientist or someone who studied much science in university or high school, you no doubt have some familiarity with biology and the life sciences. In this age of the Internet and pervasive communication, it is difficult not to at least passively soak up some knowledge about genetics, evolution, ecology, or medicine. Here's a sample (in my own words) of what most people probably think they understand about biology and the modern life sciences:

> Genes or DNA provide the code or instructions or program or blueprint or whatever for building organisms, which are made from tiny cells that contain proteins and things that are like machines or factories, and they got that way through evolution because nature selected only the strongest organisms to survive and reproduce, which explains why they all fit perfectly into their little niches in the environment. But when humans mess up this balance of nature we make ecology unhealthy, which can lead to illnesses like cancer. Fortunately, scientists are now experimenting with molecular scissors like CRISPR to do gene-editing that will reprogram or rewire cells to switch the cancer off.

This invented statement is only meant to summarize what I believe to be fairly common opinions people have about issues in the life sciences, based on my own 20+ years of experience of teaching university courses on science and society and research in the history and philosophy of science. The ideas expressed in the made-up passage are not all entirely false or off-base, but they do suggest a less-than-firm grasp of what the terms employed really mean. Part of the problem from the non-scientist's perspective is that science, when it's not using a lot of incomprehensible mathematical equations and technical jargon, contains a lot of metaphorical language. And that's because scientists actually do use a lot of metaphors, not only when they are communicating with laypeople outside of their profession, but also when they are actually doing their science. In many cases, the metaphors actually help them to think about their research questions, how to set up their experiments, and how to interpret the results.

In the prologue to his book *How to Grow a Human* (2019), science writer Philip Ball describes his growing awareness of how much science is "driven by stories" or narratives that inform our interpretation of what the science means. The metaphors scientists use to describe the things they study in particular influence how they think about them and implicitly suggest stories or narratives through which they may be understood; and although science is commonly portrayed as providing an objective account of the world, close inspection reveals that it is replete with and reliant on metaphorical language and concepts. Because metaphors create bridges between two ostensibly dissimilar topics, this makes them powerful facilitators of analogical reasoning, which allows scientists to apply what they already know about one type of entity or process to others more novel and poorly understood. For instance, thinking of biological cells in analogy with factories has permitted scientists to apply insights into how factories are organized and operate to the structure and function of cells.

It can be challenging, however, for non-scientists or anyone unfamiliar with the precise details of how scientists use and interpret these metaphors to understand exactly what their pronouncements do and do not mean. In many cases, scientists use metaphors like "chemical bond" or "genetic code" as a kind of shorthand that covers a whole range of quite specific and well-understood ideas, phenomena, and techniques. But not always. Sometimes the scientists themselves are as uncertain as any of us what precisely the metaphors mean, even though they may find them quite useful for some purpose. But even when the scientists do understand the metaphors in quite specific ways that are well accepted within their professional community, they can be misleading and confusing to those on the outside. Just as anyone can look through a microscope without knowing how to make sense of what they see, knowing how to interpret a scientific metaphor requires a little instruction. This book aims to provide such instruction in brief and accessible language.

As Brendon Larson, a conservation biologist who has written on the relationship between metaphor and science, says: "Scientists are responsible for their metaphoric choices and citizens are responsible for learning to interpret scientific metaphors." And as the image on the cover of this book suggests, metaphors consist in the recycling of ideas from one domain of discourse to another. Frequently, ideas from domains outside of science are transferred into a scientific topic, as, for example, when the ideas of codes, information, and computer programs were injected into the field of genetics. But the ideas, once established in a scientific domain, can then be recycled and transferred back into non-scientific conversation, as when we ask whether we should edit our genomes as a means of treating disease or disability. This entrenches the idea that both the cause and solution are biological and located inside our bodies, rather than perhaps implicated in a broader network of relations extending beyond our genomes to the natural and social environments in which we live.

In this respect, metaphors are like viral vectors that carry ideas (and habits of thought) bi-directionally between science and society. And like a virus, once a metaphor has settled in (as a conventional or "dead" metaphor), we may no longer recognize we are speaking and thinking under its influence, and we may uncritically replicate and perpetuate its existence to the exclusion of other potentially more useful modes of speech and thought.

The purpose of this book is to help everyone – non-scientists and scientists too, I hope – to think more clearly about the many functions of metaphors in science, and to understand better the meaning and roles (both positive and negative) of a selection of specific metaphors drawn from the various life sciences. Think of it as a kind of booster shot for your critical thinking response system.

Chapter 1 provides a basic introduction to metaphor and to ideas about its relationship and relevance to science. Chapter 2 explains how and why a small set of rather general sorts of metaphors have been so common in science, both past and present. In Chapters 3 through 8, I discuss a select set of metaphors specific to a particular area or field in the life sciences, covering genetics, proteins, cells, evolution, ecology, and biomedicine. Each of these chapters begins with a brief account of the metaphors to be discussed, followed by a short history of why scientists began using these metaphors in the first place, explaining why they have been helpful for the investigations and theories in which they occur, after which the deficiencies and reasons why the metaphors have been criticized are discussed.

In short, we will be asking questions such as

- Are genes blueprints?
- Are cells factories?
- Are proteins machines?
- Does nature select which organisms get to survive and reproduce?
- What exactly is the tree of life and what kind of tree is it?
- Is there such a thing as the balance of nature?
- How do scientists go about editing a genome?
- Will they really be able to switch off or reprogram cancer cells?

The literature on metaphor and its relation to science is vast, and I have benefitted from a great number of authors and publications. I have tried to draw attention to those examples that I think are particularly helpful for understanding the select set of metaphors discussed in this book. I do not pretend to offer the final word on the topic of science and metaphor; this is rather a snapshot of some of the interesting topics and research to date. To those whose work I have not mentioned or have missed in my own reading, I extend my apology.

Acknowledgements

I am very grateful to a number of people who read drafts of chapters and provided helpful suggestions and corrections, including Ford Doolittle, Francisco Gomez-Holtved, Paul Handford, Andrew Inkpen, Dani Inkpen, Richard Keshen, Brendon Larson, Brigitte Nerlich, Daniel Nicholson, Michael Ruse, and Kellie White. Thanks to Professor Andrea Streit for providing the image for Figure 8.3a, and to Cape Breton University librarian Jasmine Hoover for assistance locating literature. To Kostas Kampourakis I owe the greatest debt of gratitude, for asking me to take on the project to begin with, for providing exceptional guidance and support throughout the process, and for reading the entirety of the manuscript and offering sage advice on content and organization that improved it significantly. I am also very grateful for the assistance and support provided by Katrina Halliday, the Executive Publisher for Life Sciences at Cambridge University Press, Olivia Boult, Senior Editorial Assistant for Life Sciences and Medicine, Sam Fearnley, Content Manager at Cambridge University Press, and Gary Smith for copy-editing the book. Of course I take full responsibility for any errors or deficiencies that might remain despite all the efforts of these excellent people.

application of words [that] obstructs the mind to a remarkable extent." But perhaps none expressed this attitude better (and ironically with such evocative metaphor) than Samuel Parker, a member of the Royal Society, who in 1666 described metaphors as "wanton and luxuriant phantasies climbing up into the Bed of Reason, [that] do not only defile it by unchast and illegitimate Embraces, but instead of real conceptions and notices of things impregnate the mind with nothing but Ayerie and Subventaneous Phantasmes." Metaphor is for romantic poets, not hard-nosed objective scientists. As the neuroscientist Michael Arbib and philosopher of science Mary Hesse explained, "the rise of science was accompanied by the conception of the 'ideal language' that would enable us to read off from the 'book of nature' the true science that exactly expresses reality." Eventually philosophers and even some scientists would have a change of opinion about this. But before we discuss the reasons for this reassessment, we need to clarify what metaphor is, and how it differs from simile, with which it is often confused.

What Is Metaphor?

The *Oxford English Dictionary* defines metaphor as "A figure of speech in which a word or phrase is applied to an object or action to which it is not literally applicable." There is good reason to extend the account of metaphor beyond the purely verbal to include visual metaphors, which are images (either natural or artfully constructed) that portray a thing in a way suggesting comparison with another distinct kind of thing. So metaphor involves talking about, describing, or thinking about one thing in terms typically ascribed to another quite different sort of thing. It can be quite explicit, such as when we say "Man is a wolf to man," or it may be more subtle, such as when we say "The sun sank down over the horizon." It is worth noting that what constitutes a literal application of a word or phrase will be subject to the practice of a community of speakers, and this can change over time, so that what was originally undoubtedly a metaphor – like the description of the sun sinking in the sky or the legs of a chair – may eventually be regarded as literal speech or a dead metaphor. (More will be said about this in later chapters.) In either case, whether obvious or not (live or dead), we use metaphor to say something interesting or novel about a subject by drawing an implicit comparison between two things commonly regarded as dissimilar (people and wolves or the sun and a leaky ship, for example). A metaphor usually suggests, rather than explicitly states, the features of the comparison on which one is intended to focus, so some interpretation is required on the part of the auditor. As it turns out, it is precisely because metaphors are open-ended that they can entail more comparisons than was originally intended by the speaker, which makes them powerful devices of intellectual suggestion and not just tricks for making speech pretty or entertaining.

Basically, then, metaphors allow us to compare two things in a striking and interesting fashion. When Leonard Cohen wrote the poem "A Kite is a Victim," he used metaphor to focus attention on some surprising similarities between two very dissimilar things. The poem highlights how a kite is subject to our control, by means of the string, but also puts up resistance as though it were an unwilling participant in what is for us an amusing pastime. Metaphor scholars refer to the concept of a kite here as the *target* of the metaphor (the thing to be described) and the concept of a victim as the *source* (of the ideas to be transferred to the target to make the description). The word *metaphor* actually derives from Greek, meaning "to carry over." The metaphor works by drawing on common associations we all have about victims and victimhood (the source domain) and transferring them to the concept of kites (the target domain) (Figure 1.1).

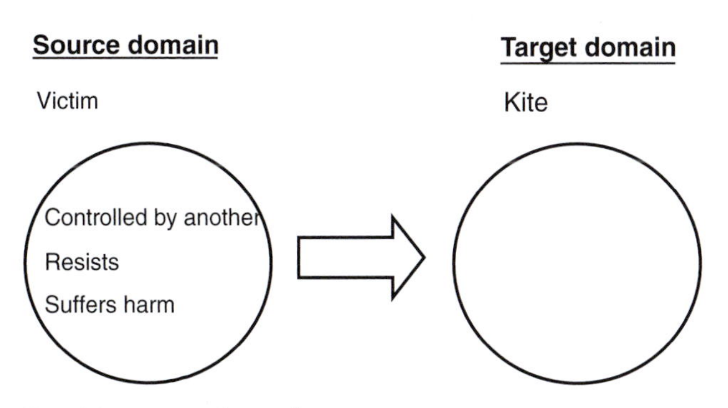

Figure 1.1 Source and target domains.

Similes also draw comparisons between things, but with the key difference that similes typically employ the words "like" or "as." For example, Cohen's song "Like a Bird on the Wire" compares his attempt to live life on his own terms ("to be free") to a variety of disparate things (including a drunk in a midnight choir and a worm on a hook). Because simile employs the terms "like" or "as," it is clear that the two things being compared are only similar in some (non-essential) properties. Metaphor, on the other hand, invites us to regard the two things as essentially identical or as two instances of the same category. A kite is not merely *similar to* or *like* a victim, the poem suggests, but actually *is* a kind of victim. And the effect is that it is difficult to look at or think about a kite in the same way after reflecting on the metaphor. The implicit assertion of identity between the relata of the source and target domains created by the metaphor explains why metaphor is such a powerful source of insight for science, as it can help us to recognize deep and non-obvious similarities and patterns between disparate things.

The Roles of Metaphor in Science

Serious attention to metaphor and its positive role in science would not occur until the early 1960s, by the philosophers Max Black and Mary Hesse. Black argued that metaphors are more than simple equivalents or substitutes for similes because the comparisons they make possible between the two domains are open-ended and, in fact, create the similarities (or lead us to "see" them to use a common lens metaphor). Saying, for instance, that "Light is a wave" is not equivalent to saying "Light is like a wave." The metaphor, he argued, is not simply a shorthand for a set of previously recognized similarities. Black also developed an insight made earlier by the philosopher I. A. Richards that the effect of a metaphor is not uni-directional, as a transfer of ideas commonly associated with the source domain to the target domain, but that our thinking about both is changed as a result of the metaphor. When we use the metaphor "Man is a wolf," the chief effect is to make humans appear wolf-like, but at the same time and to a lesser degree perhaps, wolves become more human-like. Black called this the Interaction View of metaphor. He also commented on how many scientific ideas, such as the wave theory of light or the billiard ball model of atoms, have their origins in metaphorical language. "Every metaphor," he wrote "is the tip of a submerged model."

The connection between metaphor and the construction of scientific models and theories was taken up by Mary Hesse in 1966, who argued that scientific explanation employing new theoretical language involves the metaphorical redescription of one type of object or event in terms originally appropriate to another system or domain. Much of theoretical explanation in the sciences, she pointed out, relies on analogical reasoning, whereby the scientist, upon recognizing some similarity between two separate systems, uses her knowledge of one familiar system to make inferences about another less familiar and more puzzling system. Metaphors, she noted, are an excellent facilitator of analogical reasoning. When we describe one system metaphorically in the terms appropriate to another (e.g., describing light as a wave, or gas molecules as little billiard balls) three types of analogical relations are established between the source and target domains, which she called positive, negative, and neutral (Figure 1.2).

The positive analogies are those properties that we know to be shared by the two systems; the negative analogies involve the properties we know to be present in one but not the other, and the neutral analogies are those features we do not yet know to be positive or negative – that is, either real similarities or dissimilarities. By highlighting neutral analogies between two separate systems, metaphors inspire avenues of experiment and investigation for scientists to follow. Proper scientific explanations rely on the identification of real and deep positive analogies between the two systems, typically ones that fix on underlying structural relations like physical laws. Hesse insisted that for successful science not just any metaphor will do, as those that trade on superficial similarities (say, colour or size) are unlikely to increase our understanding or ability to control events in the world.

The most influential advocates of the thesis that metaphors play an important cognitive role in how we think about and experience the world are the cognitive linguist George Lakoff and the philosopher Mark Johnson, whose 1980 book *Metaphors We Live By* introduced what is known as the conceptual metaphor theory. According to the authors, humans rely on an extensive range of metaphors drawn from a set of basic and immediate experiences to help organize and conceptualize more abstract ideas and events. So, for instance, because we associate health with standing erect (up) and illness with lying prostrate (down), we use these experiences and our descriptions of

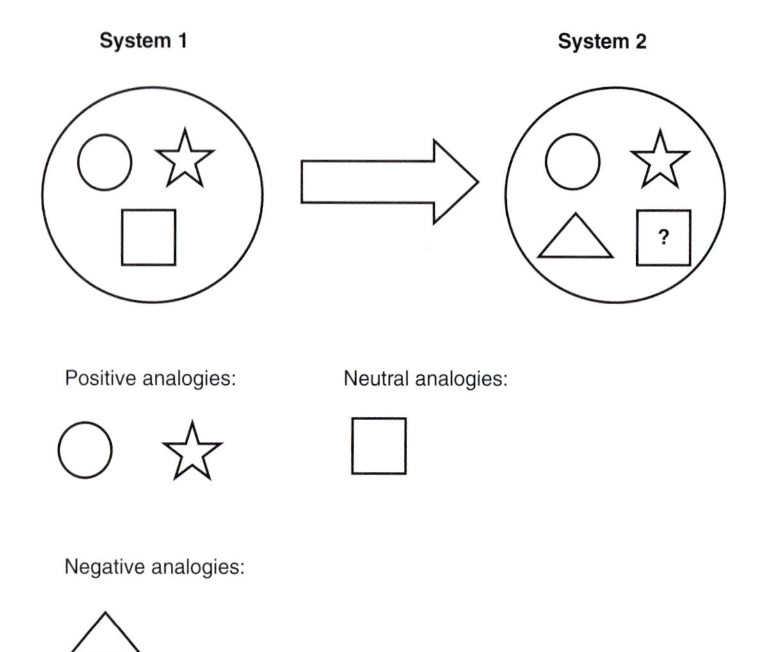

Figure 1.2 Positive, negative, and neutral analogies.

them as source domains to create and to organize our understanding of more abstract target domains, such as the concept of the future. We say, "Things are looking up!" if prospects are good. Or we describe positive economic activity as "growth" (because when a pile of objects increases in size it extends upward) or as a "healthy economy." Negative economic activity is described as "sluggish," "in recession," "depression," or an "economic slump," all suggestive of the lack of activity and uprightness associated with illness. We describe and think about the future and life generally as a journey: "No one knows where the future leads or what's on the road ahead, and sometimes there will be obstacles and crossroads at which you will have to make tough decisions, etc."

One of their most interesting assertions is that not only are metaphors not inconsequential *façons de parler* but essential components of how we describe and make sense of the world, they also, as the title of their book proclaims, shape the very way we live and experience the world. Because we employ the conceptual metaphor "life is a journey," we actually experience it that way, and at least among most members of the Western world, we experience the future as though it were spatially situated in front of us and the past behind us, and think of ourselves as travelling toward it as if it were a destination.

So far we have seen that metaphors play at least two important roles in science: (1) a *heuristic* role, suggesting analogical models or hypotheses to be explored in order to discover important unifying similarities or patterns in nature (functioning in what philosophers of science have traditionally called *the context of discovery* where creativity is all that matters and "anything goes"); and (2) a *cognitive* role in the development of explanations that increase our understanding of nature and its mechanisms (functioning in *the context of justification*, where logic, experiment, and careful attention to evidence and its interpretation are supposed to rule). To this list we may add a third role: (3) as a *pedagogical* or *rhetorical* device in the communication between scientists and non-scientists and students. This is the function most frequently and grudgingly conceded to metaphor by those who wish to defend the image of science as the purely objective and factual account of how the world really is. Sure, these people will say, scientists engage in metaphor, but that's only to dumb down the highly technical and difficult jargon-laden language that they use to make sense of the world. They do this so that non-scientists and students can get some basic understanding of what the scientists are *really* saying and thinking when they are in the lab or actually doing the science.

It is unquestionably true that scientists and science journalists do employ all kinds of creative metaphors for the strictly rhetorical purpose of helping them communicate difficult and unfamiliar ideas to an audience of non-scientists. But as we shall see throughout this book, the heuristic and cognitive roles metaphors play in science cannot be denied without grossly mischaracterizing the actual process of scientific practice and the ultimate product – namely the knowledge, theory, explanation, and understanding that science

> The hope for any metaphor in science is that it may bring otherwise unfamiliar subjects to life, make connections not otherwise apparent, and stimulate fruitful inquiry. A danger is that a metaphor can restrict rather than expand research horizons.
>
> The problem … is not so much that a metaphor is wrong but that it is misleading: It encourages the interpretation of a partial view as the whole truth or the attribution of too much importance to the view provided by one metaphor as opposed to the different insights provided by a plurality of them.

It is important to recognize that every metaphor provides at best a *partial* and *selective* perspective on reality, and that it may be important to adopt several different metaphors if we want a more complete (more objective) understanding; just as we typically attempt to view an object, like a statue, or an issue of debate *metaphorically* "from all sides." We can think of metaphors as providing a path or map forward, through a tangled jungle of unfamiliar territory (to use yet another meta-metaphor), but like any map they may not lead us to where we ultimately want to go; and if we do not exercise due caution, we can be easily misled by them.

Metaphor's Broader Impact Beyond Science

In the chapters to follow, we will look closely at examples of metaphors that have been highly conducive to advancement in various branches of the life sciences, some that have been less so, some that are a mix, and some about which the jury is still out. We will also consider the equally important question of the impact that metaphors can have on broader society via science communication, or their rhetorical-persuasive effects. As the population geneticist Richard Lewontin has said, in addition to helping us to understand the world and to manipulate it to our advantage, through the accounts it gives of the way things are, science also works to *legitimate* and support various political, economic, and social ideologies. Because metaphors work by drawing on common beliefs and attitudes associated with a source domain, implicit values and value judgments can also be transferred to the target concept, but often implicitly and therefore escaping critical scrutiny. Empirical studies indicate that the choice of metaphor used to frame a problem (e.g., describing crime as either a beast or a virus) can significantly influence people's reactions toward it and the solutions they consider appropriate. This provides further reason to pay close attention to scientific metaphors, for as biologists Cynthia Taylor and Bryan M. Dewsbury write, "the metaphors we rely upon may uphold and reinforce outdated scientific paradigms, contributing to public misunderstandings about complex scientific issues." And as the professor of science, language, and society Brigitte Nerlich and her colleagues explain:

> Metaphors can be used to highlight and hide or foreground and background issues for specific purposes. It can attract attention, increase funding, excite people, encourage them to accept a new technology and so on. Clearly, this process has political and ethical implications that need to be understood and discussed.

Some metaphors, designated promotional metaphors by the sociologist Dorothy Nelkin, are employed not so much to highlight some feature of reality that might otherwise remain obscure, but for the rhetorical purpose of attracting public and private investor support for what has become the expensive business of running complex laboratories that depend on high-tech equipment and highly trained technicians. At least some of the talk of the genome as the "blueprint" or "book of life" used in the 1980s and 1990s by those advocating for the Human Genome Project could be described as promotional metaphor.

This highlights that the use of a metaphor to describe something is not merely descriptive, it is also a performative act whereby the speaker is prescribing – either implicitly or explicitly – that it is right to think about the thing in accordance with the metaphor and to feel emotionally toward it in accordance with all the normative values with which it is commonly associated. Because metaphors may, albeit unintentionally, conjure up associations that are socially or politically problematic (as we will see in later chapters), many scientists are becoming increasingly aware that they have an extra duty to reflect on how the language they use to describe their topic of research will be interpreted outside of their professional community and in broader society.

Miscommunication Between Scientists and Non-scientists

In general, there are two ways the community of scientists and the general public can be in disagreement or confusion about some term or phrase in scientific currency. The scientists may interpret the language in question as an instance of metaphor, while the non-scientist takes it literally; or the scientists take the language literally and the non-scientists interpret it as metaphor. In the first scenario, a scientist when talking with a layperson about some particular word or phrase (for instance, *selfish genes*) may say, "Oh, but that's only metaphor! You can't take it literally." And the scientist may explain that when they use the term, they understand it as a kind of shorthand for some other set of ideas that could, with less brevity, be expressed in literal language (see Chapter 6). In the second scenario, the non-scientist may suggest that a particular scientific term or phrase is a metaphor (e.g., that all organisms are composed of *cells*, or the whole topic of *signal transduction* in cell communication), to which the scientist might respond, "No, it's not!", because for them these terms and phrases – although they may concede were originally introduced as metaphors – have now become so specifically associated with well-understood phenomena, processes, and/or techniques (positive analogies only) that they are now "dead" metaphors, which is to say that the community of scientific specialists construes them quite literally as technical terms distanced from their original metaphorical connotations.

It is important to reiterate that there is no purely objective and eternal answer to the question of whether a particular term or phrase is literal or metaphorical. This is because whether the use of a word or phrase is literal or metaphorical is always indexed to the conventional norms of language use of a particular community of language speakers at a particular time. What was metaphorical at one time (e.g., talk of chair legs or biological cells) may become literal over time. It is also important to note that the issue of whether a particular use of a word or phrase is literal or not is distinct from whether a statement in which it appears is objectively (i.e., independently of humans) true or false. It may be a literal use of language to say that the chair I am currently sitting on has four legs, but that is not to say that it is an objective truth about reality as it exists independently of us that chairs have legs. It is a truth that most humans would assent to, but obviously objective reality has no conception of and says nothing about chairs or the concept of chairs. And likewise, it may be a literal use of language to say that organisms are made of cells, but whether that is an objective truth about reality as it is independent of the human mind is another matter, and one about which I will say no more here.

Summary

Metaphor plays several well-documented roles in science: (1) **rhetorical** (this involves the pedagogical use of metaphor in science education or in communication between scientists and non-scientists); (2) **heuristic** (facilitating discovery, creation of novel hypotheses and paradigms), where the idea is that the metaphors are initially useful in the creative process of doing science, in the context of discovery, but that they may eventually be replaced by more precise and rigorous literal language; (3) **cognitive** (enabling analogical reasoning, explanation, and scientific understanding), the claim being that metaphor can and does play a legitimate role not only in the process of science but also in the final product, namely scientific explanation, understanding, and knowledge: The metaphors do not necessarily drop out or get replaced by more rigorous and literal terminology. I suggest that we should recognize a fourth role for metaphor in science: (4) as a kind of **technological instrument** that changes not only the way we think about or understand the world, but also leads to real material change in the very nature of the thing to which the metaphor is applied. I will discuss several examples of this fourth role in later chapters, using the cell factory and cell rewiring and reprogramming metaphors as examples.

2 Background Metaphors

Agents, Machines, and Information

Given the wide range of possibilities to draw from, one might expect the metaphors being used in the life sciences to come from a wide variety of source domains. After all, if you're trying to describe an organism and understand how it works, for instance, you could in theory compare it to anything. But as a matter of fact, the metaphors one tends to find in the life sciences fall into three broad categories: *agents, machines,* and *information*. I will refer to these broad categories as background metaphors. All three involve teleological thinking – that is, the assumption that things are (or that it is at least a helpful heuristic to suppose they are) either designed to fulfill certain functions or have plans of their own they are attempting to achieve. We will also look at a smaller number of metaphors drawing on natural objects as the source domain, but the majority to be covered in this book will fall into the three chief background metaphor categories of agents, machines, and information.

Agent Metaphors

By an agent we mean an entity with deliberate and goal-directed behaviour, or as we say, agency. Humans are of course the archetype. The history of human civilization across the globe reveals a strong and early tendency to engage in the teleological thinking characteristic of animism and anthropomorphism, whereby human consciousness and efforts to fulfill various desires are projected onto other living and non-living things, from other animals to trees and forests, rivers, oceans, clouds, storms, the stars, the planet, and the universe as a whole. By assuming something is alive and ascribing intentions to it, people have been able to make some sense of events. This is what the philosopher Daniel Dennett has called the intentional stance. In doing so we make what is unfamiliar familiar: Why did the sea suddenly get so rough and drown all my relatives? It must be angry at us for some reason. Teleological thinking is one way to identify or to impose an "order of nature." Whether it is factually correct to do so is another matter.

It's understandable that we tend to anthropomorphize something the closer it is to being human, in which case the number of similarities or positive analogies is large. We quite naturally assume, for instance, that other mammals (apes, cats, dogs) have minds and thoughts, emotions, and desires like us, all of which can help to explain their behaviour. And it's very difficult to resist taking the intentional stance toward many species of bird, especially intelligent ones like crows and parrots. We may be a little more hesitant to do so with fish, ants, worms, or microbes. But as we will see in later chapters, despite our having gone through the scientific revolution and its strict invocation against excessive anthropomorphizing, modern scientists still frequently talk metaphorically of genes, proteins, cells, and nature itself as if they were conscious agents. Darwin's theory of natural selection is itself based on the metaphorical portrayal of nature or the environment as selecting or favouring some variations in traits over others, in analogy with how animal and plant breeders select for particular traits in their stock to create new and improved varieties of horse or apple, for instance.

Agent metaphors are especially popular, however, when describing function. For instance, proteins or complexes of proteins are frequently said to "cooperate" and to "recruit" one another to fulfill specific tasks. Damaged cells in our bodies are said to "commit suicide" altruistically for the benefit of the body as a whole, and the body itself is not uncommonly spoken of as a "society of cells." Such agential language helps to frame the activity of the entity in question in a larger context or "social network" of relations. These agential metaphors help describe the behaviour or activity of the things in question in ways that are readily intelligible and familiar to us humans. But for explaining *how* proteins or other agents manage to perform the function in question, scientists typically turn to another background metaphor: machines.

Machine Metaphors

The standard historical account of the scientific revolution emphasizes how the prior anthropomorphic, animistic, or organicist worldview associated with Aristotelian philosophy was gradually replaced by the new mechanical philosophy and its reliance on machine metaphors. The promotion of machine metaphors and analogies by Descartes and other early modern thinkers, in their attempts to understand the functioning of living bodies, reveals the influence of technological development on the scientific revolution. It may be no coincidence that this important shift in thinking occurred at a time when curious minds were increasingly surrounded by new technological devices such as hydraulic systems, mechanical clocks, and gear-driven automatons made to look and move like humans and other animals. Once people got good at building complicated machines composed of lots of different parts, they could then get good at taking other natural objects that were complicated apart (conceptually and materially) to figure out how they work (Figure 2.1). Technological innovations like hydraulic cylinders, mechanical clocks, and electric telegraph systems became the source of hypothetical models and analogies to help tease out how living organisms might operate. Like the use of agent metaphors before, machine metaphors were a useful heuristic that facilitated analogical reasoning. In contrast to complaints that agent metaphors lead to anthropomorphism, the use of machine metaphors may be characterized as *technomorphic* and potentially resulting in the parallel sin of *technomorphism*, the projection of features characteristic of machines and other technology onto entities and processes that are natural rather than human artifacts.

The rise of the mechanistic worldview in the sixteenth and seventeenth centuries has led to the ubiquitous talk of mechanisms in current scientific practice. This allows natural objects and processes to be conceptually decomposed into simpler and more intelligible parts and processes. As the population geneticist Richard Lewontin writes:

> It seems impossible to do science without metaphors. Biology since the 17th century has been a working out of Descarte's [sic] original metaphor of the organism as machine. But the use of metaphor carries with it the consequence that we construct our view of the world, and formulate

(a)

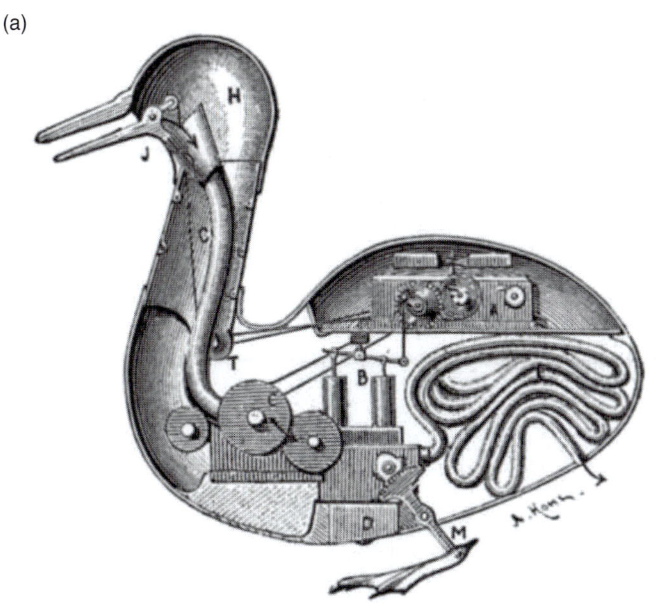

Figure 2.1 (a) A depiction of Jacques Vaucanson's mechanical duck from the eighteenth century (public domain). (b) Man as machine: Fritz Kahn's *Der Mensch als Industriepalast* (Kosmus, 1926) (public domain).

> our methods for its analysis, as if the metaphor were the thing itself. The organism has long since ceased to be viewed *like* a machine and is said to *be* a machine.

Indeed, much of what goes on in the life sciences today can be described as the search for causal mechanisms to explain how or why living things behave the ways they do. In the most general sense, mechanisms can be characterized as "entities and activities organized such that they are productive of

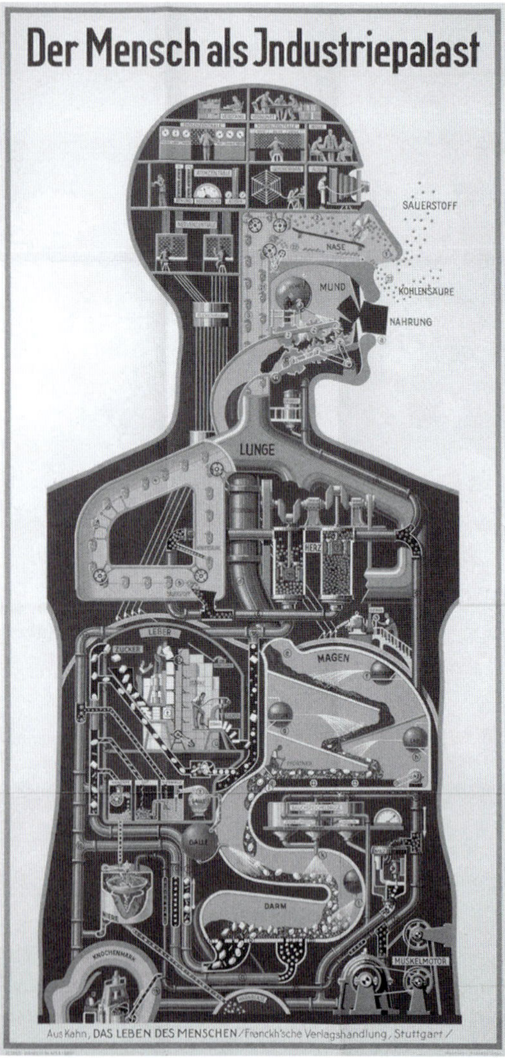

Figure 2.1 (cont.)

regular changes from start or set-up to finish or termination conditions." To search for a mechanism in the present age is basically to create a causal account of how something works. So, while the notion of a mechanism is undeniably descended metaphorically from our familiarity with and understanding of machines, the meaning of the term has changed significantly. Over time the term "mechanism" has become *polysemic*, which is to say that the original meaning has branched into two distinct though related senses: one referring to the set-up of a machine, the other referring to any causal relationship in general. As mechanism-talk regarding living organisms and their component systems (organs and cells) became more common and more firmly operationalized, explicit analogies to specific kinds of human machines became less important and mechanism became a dead or conventional metaphor. The philosopher of biology Michael Ruse explains that when scientists today speak of the "mechanism of inheritance" or the "mechanism of selection," there is no suggestion of nor need to invoke imagery of any machine-like arrangement of material parts; all that is implied is a regular causal relation between an initial and final state of some system. However, as we will see in later chapters, some mechanism-talk in the life sciences does still cling closely to the traditional image of a machine as a closely defined spatio-temporal arrangement of material parts.

As technology has evolved, so too has our understanding of what machines and mechanisms can look like. The rise of the computer age beginning in the mid-twentieth century has created a whole new world of interconnected digital devices processing and exchanging not material products like cotton, spices, or precious metals as in times past, but the more nebulous commodity known as information. This has helped to create the third category of background metaphor which has so significantly shaped the modern life sciences.

Information Metaphors

The pioneer of cybernetics, Norbert Wiener, wrote, "If the seventeenth and early eighteenth centuries are the age of clocks, and the later eighteenth and the nineteenth constitute the age of steam engines, the present time is the age of communication and control." Cyberneticists like Wiener – who worked during the Second World War on guided anti-aircraft weapons systems –

agency to run the cell ("The genome is the program that creates the organism"). Alternatively phrased, the "genome is the software" and the "cell and organism are hardware."

More recently this has resulted in the talk, now common among synthetic biologists and health researchers, of genome editing, rewiring genetic circuits, and reprogramming cells as a strategy to treat or prevent disease. Scientists have come to recognize that cells are not simple computers running on hardwired circuits etched into a static motherboard. They contain complex networks of dynamic signalling pathways that are sensitive to external inputs from other cells, organs, and the external environment beyond the body. Moreover, these external signals can result in so-called epigenetic modifications to the three-dimensional "landscape" of the tightly coiled complex of DNA and histone proteins known as chromatin and its location in the cell's nuclear "architecture," all of which affects which bits of the purported genetic program are accessible to be "read" and "expressed" as proteins or other genetic "products." And yet, it is undeniable that scientists are achieving some level of success in their pursuit of this metaphorical program to rewire and reprogram cell function and behaviour. The use of recombinant DNA techniques in the late 1970s and early 1980s allowed scientists to genetically engineer bacteria (*E. coli*) and yeast (*S. cerevisiae*) to produce human insulin, effectively turning them into literal cell factories that have saved countless human lives from the ravages of diabetes. (Prior to this insulin was harvested from the pancreases of cattle and pigs.)

With the assistance of newer techniques such as CRISPR-Cas9, scientists hope to be able to make specific, predictable, and safe alterations to the genetic instructions cells use to synthesize proteins implicated in human diseases – for instance, removing a damaged gene and replacing it with a functional one or using "molecular scissors" to "cut and paste" genes, as the popular description has it, in a mixture of old-fashioned machine technology with modern computer software metaphors. Synthetic biologists like J. Craig Venter are even attempting to build from scratch a synthetic cell using these computer engineering design metaphors. In 2010 Venter and his team announced the creation of *Mycoplasma mycoides* JCVI-syn1.0 (also known as *Mycoplasma laboratorium* or "Synthia"), a bacterial cell functioning with a genome designed and synthesized in the laboratory with the assistance of digital computer technology. Although Synthia1.0 and its later version Synthia3.0 do not yet qualify as fully synthetic life forms, as they rely on a natural pre-existing cell "chassis" emptied of its own original genome, they do raise the possibility that one day metaphorical talk that "cells are computers" could become quite literal.

All this talk of information, codes, and programs inevitably invites the question: Who wrote the book/code/program of life? Most scientists will say it was written by nature itself or, more specifically, natural selection. Creationists and advocates of intelligent design theory insist that only an intelligent agent (i.e., god) could have done so. But as the historian Lily Kay noted, both of these responses take the metaphors literally, whereas she made the case that it was the people who, under the influence of the language and metaphors of codes, texts, and programs that they introduced, created this particular way of talking about and understanding the world. Which is not to say that genes are not real, and that DNA and proteins are all social constructions; nor is it to deny that this language has been immensely useful and successful in increasing our understanding and ability to control and manipulate living systems, including ourselves. But as a narrative she suggests that it is no more inevitable or inherently objectively *nature's own account* than any other human narrative one might tell. "These particular representations are historically specific and culturally contingent," she notes. They have worked well for us in some respects, but there are already calls from scientists themselves that they must be at least modified if not wholly replaced with alternative metaphors if further progress is to be made.

Language Is the Primary Tool-Box of Science

Calls for alternative metaphors reflect a recognition that there is no way to describe or talk about the natural world in its own terms because nature does not have its own inherent or objectively correct ("true") language. Language is a human device that we use for a whole host of purposes: formulating and communicating thoughts and emotions, making promises and threats, creating and telling stories, socially coordinating our behaviour, and so on. Science requires language that allows us to attain the most reliable, empirically adequate, and useful description and understanding of phenomena that range from the microscopically small (cellular, molecular, of DNA spooled like thread around histone proteins and how it coils and unwinds to allow genetic sequences to be transcribed into messenger RNA.

It seems that these non-artifact, non-machine, non-script metaphors chiefly serve to provide an organizing image or picture of how the components of the phenomenon in question stand in relation to one another. They provide a particular way of framing the subject in question and do not, in contrast to the human artifact metaphors, attempt to provide a causal explanation or mechanism of *how* the phenomena occur. Each major category of metaphor, however, brings with it assumptions about the nature of causation involved. Machine metaphors, for instance, subtly promote the assumption that the living system (whether an ecosystem, an organism, or a cell) is composed of distinct parts that retain their essential properties in isolation of the whole, in analogy to the gears and springs of a clock. Agent metaphors suggest the potential for mutual cause and effect, the actor being changed as a result of the action, in addition to the appropriateness of goal-directed narratives to describe the behaviour in question. Information metaphors are open to multiple interpretations, depending on the theory of information and information transfer assumed (the effect of information input in a computer system can be quite different and far more predictable, for instance, than the input of information in the form of a rumour into a social or political setting). And even though the natural (non-artifact) metaphors mentioned above do not appear chiefly to fulfill causal explanatory functions, the image of the specific example in question may suggest distinct causal modes: a tree suggesting progressively "upward" and persistent bifurcating effects in contrast to a waterfall suggesting a "downward" cascade of events that both separate and rejoin continually.

We turn now to discussion of some of the most common, influential (and often misleading) metaphors used in the scientific study of genes and genomes.

3 Genes and Genomes

Agents, Codes, Programs, Blueprints, and Books

This chapter deals with the thorniest and most tangled thicket of metaphors, and there is probably no other area in the life sciences whose language has received so much critical attention. A great deal has been written about the metaphors used in genetics and genomics research, and I will attempt to provide only a summary here, with few original contributions of my own.

Classical genetics (that which preceded the "molecular revolution" of the mid-twentieth century) dealt with the phenomenon of biological inheritance, where evident species- and family-level similarities between parent and offspring attest that *something* is transmitted from one generation to the next (hence the metaphor of *inheritance*). Dogs give birth to dogs, corn plants produce more corn plants, and children tend to look like their parents and close relations. By the early twentieth century the gene was identified as the agent responsible for creating and transmitting these phenotypic traits. Molecular genetics peers into the details of how this inheritance is possible by focusing on the molecular structure of DNA, the genetic material, and in particular how sequences of chemical bases (the As, Ts, Cs, and Gs of the genetic language) are rewritten or transcribed by the cell's machinery into shorter RNA messenger molecules (with thymine replaced by uracil) that are then translated into amino acid chains. These polypeptides, as they are called, when folded into various three-dimensional shapes become the proteins that carry out all the various structural and functional tasks required in the living cell. A short synopsis of the popular account goes something like this:

type of cell, tissue, or organ from which they originated. These gemmules would then be responsible for the development of the specialized cells and tissues in the next generation of organisms. Assuming that gemmules could lie dormant throughout the body, Darwin suggested this hypothesis could also explain the phenomena of wound healing and regeneration of limbs (as seen in amphibians and sea stars, for instance) or of whole organisms from mere parts (as observed in polyps or worms). It would also provide a mechanism for the inheritance of traits acquired over the lifetime of an organism, a phenomenon popularly associated with the French biologist Lamarck, but also accepted by Darwin as well.

Others rejected Darwin's hypothesis of pangenesis, as evidence against the inheritance of acquired traits accumulated and investigations with the microscope made clearer that the chromosomes – the long filaments in the cell nucleus that became apparent when stained with dye – were the carriers of the hereditary material. In 1899, the botanist Hugo De Vries dubbed the suspected units of heredity "pangenes," a term that would be abbreviated in 1909 by Wilhelm Johannsen to "gene." Johannsen recommended this shorter word because it was, in his words, "free from any hypotheses ... [and] expresses only the evident fact that ... many characteristics of the organism are specified in the gametes by means of special conditions, foundations, and determiners which are present in unique, separate, and thereby independent ways – in short, precisely what we wish to call genes." Johannsen also introduced the terms "phenotype" and "genotype" to denote the physical characteristic (e.g., flower or eye colour) and its underlying genetic basis, respectively.

For the next several decades the gene remained a hypothetical entity, but that did not stop scientists – now known as *geneticists* – from attributing to it, as Evelyn Fox Keller says, agency, autonomy, and causal primacy. What she calls the "discourse of gene action" made genes responsible for all the chief elements of growth, development, and reproduction. Genes were spoken of as vital active agents and credited with transmitting traits (such as flower or eye colour, seed shape, or straight versus curly hair) from one generation to the next, and increasingly from the 1940s on for the more specific creation of enzymes and other proteins required for cell metabolism. In short, the importance of genes was magnified to such an extent that the rest of the cell, with all its organelles and other cytoplasmic contents, were largely squeezed to the margins of the scientific field of view.

The Molecular Biological Gene (1950s to Present): Information and Codes

Genes acquired a new degree of reality and materiality with Watson and Crick's 1953 papers in which they announced the double-helix structure of the DNA molecule and its implications for understanding heredity. The four bases (adenine, thymine, guanine, and cytosine) that attach to the sugar–phosphate backbone of the long winding molecule appeared always to bind regularly in a pattern of adenine with thymine (A–T) and cytosine with guanine (C–G), leading Crick and Watson to suggest that the sequence of bases running along the DNA strand "is the code" that allows the cell to replicate and pass on "the genetical information" necessary to produce a new generation of cells or organism. A gene, they proposed, would correspond to the pattern of bases running sequentially along the complementary strands of the DNA molecule. As each chain of the double-helix carries a sequence of bases complementary to the other (A on one side must be matched by T on the other and G with C), by unwinding the chain a dividing cell has two "templates" of the DNA sequence to pass on to each of the daughter cells. Furthermore, they proposed that exposure to radiation or other injurious events could cause an alteration in the gene's sequence of bases, say an A is changed to a C or G, and such mutation could result in a novel trait emerging in the next generation, which would provide variation for natural selection to operate on to produce new species.

Watson and Crick freely admitted that while their model provided an answer to the question of how the cell's genetic material was able, in their words, "to duplicate itself," it left unexplained how it managed to "exert a highly specific influence on the cell," or in other words, what role it played in the synthesis of the proteins that performed most of the work in the cell. Over the next several years, scientists working on bacteria would begin to fill in the missing details. Once conceived as indivisible "beads on a string," Seymour Benzer's work on bacteriophages (viruses that infect bacteria) in the 1950s confirmed that genes were linear stretches of nucleotides (a nucleotide consists of a chemical base attached to a sugar and a phosphate group, the "backbone" of the DNA in a developing embryo become a muscle cell and another a nerve or kidney cell? Why aren't all the genes available to each cell active all the time? An important step toward answering these questions emerged from a rather unlikely source: single-celled bacteria. Arthur Pardee, François Jacob, and Jacques Monod studied the bacterium *E. coli* for its ability to selectively produce enzymes specific for the digestion of selected sugars. Their experiments (known popularly as the "PaJaMa" experiments in a playful take on the first syllables of the authors' surnames) revealed that the presence of a particular type of sugar (lactose) could induce the synthesis of the enzyme required for its digestion (β-galactosidase), suggesting that in addition to genes for the production of enzymes and other proteins there might also be genes involved in the regulation of gene activity.

In 1961, Jacob and Monod published a paper outlining a model for gene regulation that made an important distinction in two separate kinds of genes: "structural" genes (so called because they are implicated in the molecular organization or structure of the amino acids comprising the protein) and "regulatory" genes whose products control the activity of structural genes by repressing or promoting their expression. The model invoked several important metaphors:

> According to the strictly structural concept, the genome is considered as a mosaic of independent molecular blue-prints for the building of individual cellular constituents. In the execution of these plans, however, co-ordination is evidently of absolute survival value. The discovery of regulator and operator genes, and of repressive regulation of the activity of structural genes, reveals that the genome contains not only a series of blue-prints, but a co-ordinated program of protein synthesis and the means of controlling its execution.

The authors were proposing that the expression of certain genes (the "blue-prints" for protein synthesis) were controlled by mechanisms (the paper was titled "Genetic regulatory mechanisms for protein synthesis") that acted like an electronic switch that could be activated by signals received from outside the cell. (Jacob explained that the idea for a switch mechanism occurred to him as he watched his son manipulate the on–off switch on his toy train set to regulate its speed.) The activity of these switches they were suggesting were coordinated according to some plan or program that increased the organism's adaptive function and response to its environment. For instance, the bacterium *E. coli* preferentially metabolizes glucose over lactose, the latter being a complex sugar that must first be broken down by the enzyme β-galactosidase into its constituents, glucose and galactose, before it can be used by *E. coli* as an energy source. If both glucose and lactose are available, the bacterium will begin producing β-galactosidase only after the glucose has been consumed. But how does a single-celled bacterium lacking a brain or even neurons "know" when to begin producing the correct enzyme only when needed?

The model (known as the lac-operon model) proposed that in the absence of lactose, the (structural) gene or "blue-print" for β-galactosidase is repressed or "switched off" by the product of a regulator gene (either a protein or an RNA molecule, they weren't sure which) that prevents the structural gene from being activated by binding to a region of DNA adjacent to it, a site they called an "operator." However, if the concentration of lactose reaches a sufficient level, it removes the repressor that is preventing the cell machinery for synthesizing protein from accessing the enzyme gene, thereby acting as an inducer for the expression of the enzyme required to digest it. The presence of lactose "switches off" the repressor that prevents synthesis of the enzyme required for its metabolism, or equivalently it "switches on" the gene for the required enzyme. Jacob had also been studying a similar phenomenon in the lambda phage, a virus that infects *E. coli*. In this case, the virus appeared to choose between two reproductive strategies: lysis (which releases the virus particles by breaking down the bacterial cell wall) or lysogeny (whereby the viral DNA becomes integrated into the host chromosome or forms a separate circular unit known as a replicon that is duplicated each time the bacterium divides). They proposed a similar mechanism was at work here. Experimental evidence suggested that several structural genes involved in a common metabolic pathway may be controlled by a common operator and repressor, forming a unit the authors named an operon (Figure 3.2).

Monod and Jacob also proposed in this paper the existence of a short-lived "messenger RNA," a complementary single-stranded copy of the DNA gene that acted as a "cytoplasmic transcript" carrying the instructions to the ribosome, where amino acids are conjoined to form proteins. The regulator gene would prevent the "transcription" of the structural gene into the RNA messenger, and thereby control expression of the structural gene in question.

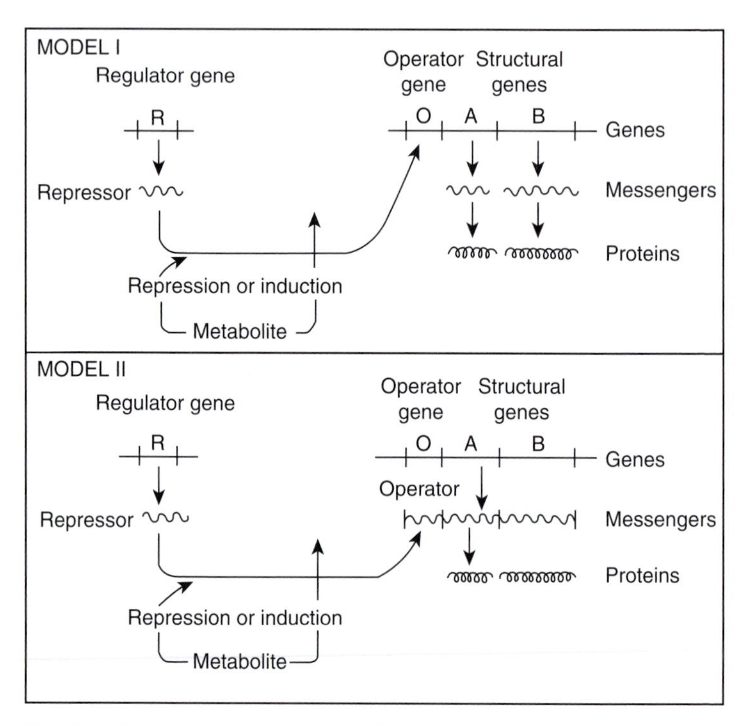

Figure 3.2 The lac operon model (reprinted from *Journal of Molecular Biology*, 3(3), Jacob, F. and Monod, J., Genetic regulatory mechanisms in the synthesis of proteins, 318–356, June 1, 1961, with permission from Elsevier).

Monod and Jacob went on to note that "the fundamental problem of chemical physiology and of embryology" is the question of why the tissue cells of higher animals do not express all the time all the genetic blueprints for proteins stored in their genomes. Protein synthesis must occur in a strictly coordinated fashion in order for a properly functioning organism to develop. And in a mature organism, a breakdown in this coordination can result in disease such as malignant cancer. For this reason, they insisted that the genome must consist not only of a series of individual blueprints, but a program to control their execution.

While it required some modification to accommodate the greater complexity of animal development and physiology, the basic idea of gene regulation founded on research on bacteria has proven extremely powerful, and the language of genetic blueprints, programs, circuits, networks, and switches now dominates scientific understanding of these events. Monod would further display and expand the computer and cybernetic roots of this approach in a chapter of his popular book *Chance and Necessity* (1970), titled "Microscopic Cybernetics." Appeal to positive and negative feedback loops (corresponding to inductive and repressive gene regulation, respectively) permitted him to explain the apparently purposive or teleological behaviour of cells (and consequently of humans) on purely mechanistic principles, without invoking any mysterious vital force or supernatural designer. Indeed, Jacob later quipped that the concept of the program had made an honest woman of teleology, earlier described by Haldane as the biologist's mistress.

The concept of a computer program was powerful because it evoked not only the idea of stored information but also of active agency, for it is commonly supposed that, if the computer is just an inert collection of physical components, it is the program that "runs" it or makes it do anything at all. But as Evelyn Fox Keller has pointed out, it is not at all clear what exactly the genetic program is, or on what it acts. DNA is an incredibly inert substance that initiates no activity on its own, but is acted upon by a system of proteins and RNA molecules. If genes are the blueprints for protein synthesis, then what does the developmental program that controls and regulates their activation and repression consist of? Protein synthesis takes place within the cellular environment that includes a system of organelles such as ribosomes, the endoplasmic reticulum, the Golgi apparatus, and is initiated in response to signals and messengers, such as transcription factors, that originate from the larger environment outside of the cell and regulate the transcription of DNA into messenger RNA. So what exactly is the program? And what "runs" it? Does DNA do the running, or is DNA part of the system being "run"? In all likelihood these are not even the right questions to ask, but the metaphors compel us to ask them.

Critical Analysis of the Metaphors

The time has come to make a critical assessment of these metaphors. What contribution has each made to scientific understanding? Do they continue to be useful? Recall that metaphors are often the seed around which scientific hypotheses crystallize. The positive analogies on which they rest suggest to researchers other possible similarities to look for (the neutral analogies). If investigation turns up more positive analogies than negative, the metaphor proves itself to offer important insight into nature, and may in time even come to be regarded as a conventional or dead metaphor, or perhaps even a literal description. So how have these metaphors stood up over the years as scientific understanding of genes and genomes has progressed?

Are Genes Agents?

Let's begin with the genes as agents metaphors. For a long time, geneticists had a way of speaking of "the gene for" some phenotypic trait, such as the gene for white flower colour, the gene for wrinkled pea coat, or the gene for cystic fibrosis. This way of talking is actually a shorthand used by geneticists to mean something like, the *allele* (i.e., the version of a specific gene associated with some trait) that is implicated in the development of a particular phenotype, such as seed shape, flower colour, or a disease condition. This gives the impression that the allele in question is "the cause of" trait *X*, or in other words, that the presence of allele *x* is a sufficient condition for the organism to exhibit phenotypic trait *X*. In fact, this is never the case. Genes (alleles) are never sufficient on their own for any trait, and geneticists have always understood this. We have already noted that DNA is not self-replicating and that DNA/genes do not "make proteins." Genes are only one component of a much broader *developmental system* that includes the entire cell, the rest of the organism if we're dealing with a multicellular organism, and the broader environment (shoot a cell out into the vacuum of space and see how little happens). Only the entire system so described is sufficient for the presence of a phenotypic trait to exist.

However, a gene may be a *necessary* condition for that trait to arise within that system. And it is in fact by discovering that certain alleles are necessary for particular traits (such as white flower colour or the absence of cystic fibrosis) that geneticists have been able to identify that these traits are under genetic influence. A standard method of doing this has been by disrupting a normal developmental system (by "knocking out" or "silencing" a gene) and observing the consequences. If it turns out that disrupting this bit of the organism's genome results in it no longer having a white flower, then that bit of the genome is labelled "the gene for white flower colour." Or if it is discovered that people with an abnormal (speaking statistically as what is most common within the relevant population) genetic sequence in this particular locus of this particular chromosome have a disease such as cystic fibrosis, then that bit of DNA becomes known as "the gene for CF." But of course this is a gross simplification of the very complex details. However, so long as geneticists understand the limitations of talking in this way among themselves, there is not much reason to criticize it. The problem arises when they speak to non-geneticists in this way, or non-scientists listen to them talking this way, and assume therefore that talk of a "gene for" a trait is an accurate account of reality. One simple way of correcting this language is to say that genes are "difference makers," in the sense that a difference in the allele can make a difference in the phenotype that is ultimately expressed by the whole developmental system. (Which is actually how early geneticists like T. H. Morgan, who pioneered fruit fly (*Drosophila*) genetics, conceived of genes.)

There is another important wrinkle in the simplistic assumption that there is a one-to-one mapping between genotype and phenotype – that is, that there is one gene responsible for each phenotypic trait. If we think more carefully about what it is that genes supposedly "do," we will recall that they provide the instructions or blueprint for building specific proteins (we will deal with the blueprint metaphor in a moment). But most of the things people think of when they hear talk of phenotypic traits are complex phenomena like the colour of hair, eyes, or skin, or disease or other conditions of poor health. A single protein, on the other hand, is seldom (though there are rare exceptions) sufficient on its own for such observable traits. Most of these very visible traits that attract our human attention are *polygenic*, which means they are the result of a combination of different proteins acting in concert. Moreover, most proteins (perhaps even all) are *pleiotropic*, meaning that they are involved in several functions within one or several distinct types of cells or tissues in the

metaphors of chromatin "landscape" and nuclear "architecture" to describe its dynamic arrangement and location within the cell's nucleus. (The dynamic response to these signals is called chromatin "remodelling.") Some of these signals are molecules such as methyl groups that bind to the histones and DNA, "marking" and "tagging" them in ways that alter chromatin conformation and activity. These alterations are implicated in the temporarily hereditary but non-mutational effects that have recently gained so much attention under the label "epigenetics." This is very different from any human code ever devised and reinforces why one needs to be cautious of simplistic talk that puts genes in control of the life of the organism.

Are Genes Blueprints?

Once scientists began to think of genes as containing the instructions for protein synthesis, it was not unnatural to think of them as providing a "blueprint" for the protein. A protein is after all a three-dimensional spatial arrangement of modular units, and in fact another popular metaphor is that proteins are made of amino acid "Lego blocks." Blueprints provide instructions for a completed three-dimensional structure in a lower dimensionality (two dimensions if drawn on a sheet or roll of paper, one dimension if encoded in a sequence of DNA bases). So here were some important positive analogies in the metaphor's favour. But beyond that, the negative analogies seem to outweigh quite heavily. For one, as mentioned above, the gene is not on its own sufficient for building the protein, nor does it contain all of the relevant "information" for the construction. Much of this "information" comes in the form of physical and chemical laws and the intra- and extracellular environmental conditions under which the chemicals that will interact to build the protein operate, and these things are not encoded in the DNA.

Sometimes the implication is that the complete set of genes, the genome, provides a blueprint for the entire organism. This is inadequate and misleading for the same reasons, only much more so. The development of an organism from the fertilized egg, which involves not only the expression of genes but also the timing and amount of their expression, is very sensitive to environmental conditions – both external and within the womb for mammals – and to stochastic events or random "noise" regarding how many signalling molecules and transcription factors are present in the cell, and how these complex signalling, metabolic, and genetic pathways play out. With so many molecular factors and interactions involved, it is hardly surprising that, although the genes can be thought of as laying down certain rules for how a game of football or hockey is to be played, each particular game will involve many unique paths and events, even if they all conclude with the predictable outcome that one team will win or it will end in a tie. My own preferred alternative metaphor to capture the stochasticity or "noisiness" of the internal cellular environment in which development occurs is "molecular weather." There are patterns to be sure, but each day is slightly different.

Others have attempted to replace the blueprint metaphor with one that gives greater recognition to the sensitivity to the environment and the contingency of events involved. The genome, the physiologist Denis Noble suggests, is more like a musical score that is interpreted by the members of the orchestra to produce a unique performance. Others, like a recent blogger for the magazine *Scientific American*, have compared it to a recipe:

> DNA is not a blueprint: it's a recipe coding for thousands of different proteins that interact with each other and with the environment, just like the ingredients of a cake in an oven ... If we are a slow-baking cake, the world surrounding us is a capricious oven, changing every minute. Science can peek through the glass and check if something looks funny inside the oven, but it cannot predict what our life, experience and luck will bring tomorrow and the day after tomorrow.

In truth, it is doubtful that the blueprint metaphor really ever qualified as even an important heuristic for thinking about and testing new ideas about genes and protein synthesis. When biologists attempt to unravel the logic of events involved in the development of an organism, they speak more specifically of "gene regulatory networks," an approach that recognizes development is far more complicated than reading a blueprint. In short, the blueprint metaphor seems largely to fulfill a rhetorical or communicative role, which is not to say it was never important nor wielded significant clout.

Is the Genome a Program?

Closely associated with the blueprint metaphor, however, is the metaphor that the genome provides a program for organismal development that regulates or controls the timing and rate of expression of the genes. The program metaphor emerged in the historical context or mid-twentieth century *zeitgeist* of cybernetic theory and the development of electronic computers. This plays to the same intuition that motivates the blueprint metaphor – that the genome encodes some kind of information or instructions for development – but supplements the static image of a sheet of paper with the dynamism and agency of a computer program to actually *do* something. All other things being equal (the environment construed broadly), it is true that genes do have a major effect on the outcome of developmental events. This becomes apparent through experimental intervention into the expression and timing of certain genes; "knocking out" (or "dialling down") the expression of particular genes at particular times often has drastic and predictable consequences. This enforces the intuition that there is some kind of logical pattern of events "encoded" in the genome. The problem is, that by assuming "all other things being equal" it has been easy to downplay the significance of those other factors, and thereby to overstate the positive analogies between a computer program and the genome.

The population geneticist Richard Lewontin, a longstanding critic of the exaggeration of the importance of genetics for development and behaviour and the metaphors that encourage it, writes, "when biologists speak of genes as 'computer programs,' they erroneously suppose that all the organism's attributes are prefigured in its genes and all that is required for a fixed output is for the *enter* key to be pressed." While it may be debatable how many working biologists believe this is actually true today, it is unarguably true of many non-scientists who have consumed a steady diet of gene-centric metaphors in the press for many years.

As Lewontin, and many others, have explained, the fertilized egg does not "compute" the adult organism from a genetic program. Development is the result of the inextricably entwined contributions of genetics, environment, and stochastic or random circumstances and events. Attempts to unravel the relative importance or contribution of genetics from environment (heritability studies), Lewontin suggests, are guided by the flawed analogy that the organism is a brick wall to which each contributes by adding its share of individual bricks. The better analogy, Lewontin offers, is that one factor mixes the mortar which the other uses to cement the bricks into place. Both "workers" are necessary and because they fulfill different roles, it is no easy task to measure who contributes more to the final result. Chance ensures that each wall is ever so slightly different even if the same workers always work together and follow the same pattern of construction. Lewontin has suggested that modern science has placed so much emphasis on DNA and the genome as the controlling agent that gives orders to the other components in the cell because modern societies tend to value intellectual effort over manual labour.

A key problem with all the gene agency, blueprint, and program metaphors is that they tend to promote what Kostas Kampourakis calls genetic fatalism: the unfortunate trio consisting of genetic determinism, genetic reductionism, and genetic essentialism. Much has been written by scientists, philosophers, and other social scientists about the pernicious roots and consequences of the idea that "DNA is destiny," and so I will not say much more about it here than the following. While of course genes and genetics are often an important component of human health and behaviour (though mostly only by allowing us to have the complex bodies and brains we have), they are only one factor and always subject to the influence of the external and internal environments in which they perform. The stochastic nature of gene expression means that rather than being the rulers of our fate, they are more like highly impressionable committee members that must work collaboratively with others to get things done. I turn now to our final key metaphor: The Book of Life.

Is the Genome the Book of Life?

If an individual gene carries the information, the instructions for assembling a protein, then the entire set of genes in an organism, its genome, must be an instruction manual or book. The description of the genome, especially our own, as the "Book of Life" has become ubiquitous since the announcement of the completion of the draft sequence of the human genome in 2000 (although its employment goes back at least to 1967, when the molecular biologist Robert Sinsheimer published a book with that title). The popularity of this metaphor is an indication of its proper function, which is not really as a hypothesis or model to provide scientists with insight into the nature and function of the genome, but as a rhetorical device to help them communicate with the broader public about the significance of the project. It is what Dorothy Nelkin called a promotional metaphor. The historian of science Lily Kay noted that metaphors "like the information and code metaphors, are exceptionally potent due to the richness of their symbolisms, their synchronic and diachronic linkages, and their scientific and cultural valences." "The Book of Life" has rich cultural and religious connotations extending back into antiquity, and brings to the scientific project a sense of awe and veneration. It proclaims that the project is not just nerdy science, but a transcendent and noble philosophical mission to gain self-knowledge. That the metaphor serves a largely promotional function is reflected in the language used to justify the budget (which in total was just under US$3 billion) to the US Congress. A passage repeated over multiple years of annual budget justifications attempted to explain the great value the public was getting for its investment.

> "The Book of Life," as some have termed the human genome, is actually three books: It is a history book that tells the narrative of the human species' journey through time. It is a shop manual that provides the parts list, and an incredibly detailed blueprint for building every human cell. And finally, it is a transformative textbook of medicine that provides insights, giving healthcare providers immense new power to treat, prevent and cure disease.

The promise that sequencing all the As, Ts, Cs, and Gs in "the" human genome would lead to a revolution in therapeutic and preventive medicine was a key selling point to attain public support. (In fact, what was sequenced in the publicly funded project was a combination of genomes from several anonymous humans; J. Craig Venter, the director of the private project that competed to complete the genome first, it was later revealed, provided the sample genome sequenced by his team.)

But Francis Collins, the American geneticist in charge of the publicly funded project, and himself a devout Christian, spoke of the Book of Life metaphor with some circumspection:

> I think it is a fairly decent analogy that the genome is our instruction book. It is maybe a bit of a grandiose statement to call it the "Book of Life," 'cause there's a lot more to life than this biological set of parts. But it does sort of bring up to mind an analogy that's pretty decent. It is, after all, a book that's written in a simple, linear fashion. It only has four letters

> instead of 26, and it's a book that we will take a long time to try to understand. It's a very large book. It's a very mysterious book. It's also basically important to think of this as not more than it is. It is, after all, a set of instructions; but it does not really tell us what being human is – nor, would I argue, it ever will.

In other words, even if it is accurate to think of genes as blueprints or instructions for protein synthesis, it adds little of actual scientific value to describe the genome as a "book." But it has real persuasive power so long as our culture retains an appreciation for books, if not necessarily veneration of one in particular. The significance of the Book of Life and program metaphors will no doubt continue to exert influence on the continued development of research, especially in light of the recent CRISPR technology ubiquitously described as a "gene-editing" tool with the power of "rewriting genomes." More will be said about CRISPR and synthetic biology's influence in biomedicine in Chapter 8.

Why did scientists place so much emphasis on DNA and genes as the "master molecule" that determines everything about life? Perhaps because in the narrow context of their specific research questions and with the tools available, it *was* the most significant component. And because genes were the focus of their investigations, it was easy to ignore or background all the other significant factors. (As the old saying goes, "To a person with a hammer, everything looks like a nail.") This is understandable and not so dangerous so long as those using the language understand that the metaphors are partial and do not reflect the whole story. But we should consider metaphors like hypotheses: They are provisional, partial, relatively useful within specific research contexts, subject to revision, replacement, or rejection. And it can be particularly problematic when these metaphors and this way of talking about life move beyond the laboratory, and non-scientists (journalists and regular folk) start using them without recognition that they are highly selective and partial reflections of a more complex reality.

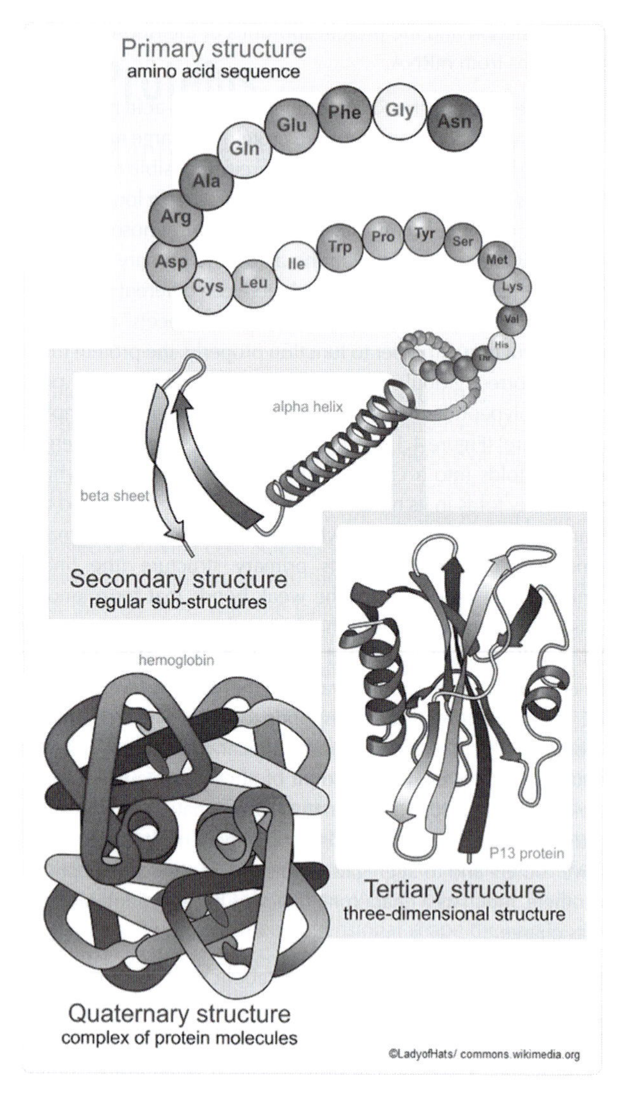

Figure 4.1 Protein structure (Creative Commons licence).

modified by the addition or subtraction of a high-energy molecule such as phosphate (called phosphorylation and dephosphorylation, respectively). Throughout all these activities the enzyme itself is left unchanged and therefore free to react with more molecules of the same sort, which makes them highly efficient catalysts of chemical reactions.

Some proteins are involved in the regulation of cell and organ function (e.g., growth hormones), acting as "signals" between cells, or as components of intracellular "signalling pathways." These include membrane-bound "receptors" of incoming signals or other components of complex signalling "cascades" by which the information relayed by an incoming signal is delivered to the nuclear DNA. In some cases, once a signal (which can be a protein, a peptide, a lipid, or a gas molecule) binds to a receptor on the membrane of the cell surface, a number of changes may be triggered in the shape or activity of closely associated proteins, creating what are known as "second messengers." In activating this conformational change in the receptor protein, the signal is said to be "transduced" across the cell membrane, invoking a metaphor from electrical engineering. This transduction of the signal to the interior of the cell can lead to the activation of a class of enzymes called kinases that act like "switches" that activate other proteins by transferring to them a high-energy phosphate molecule. This cascade of events (like a waterfall) may eventually result in the activation of a transcription factor, a class of proteins that help to "switch genes on or off" by binding to specific DNA sequences. Once bound to the DNA, they will either promote or inhibit the transcription of a coding region of bases into messenger RNA. In a display of the complicated circuit-logic of the cell, because transcription factors are proteins, the genes that encode them are themselves subject to regulation by transcription factors, and in some cases they even regulate themselves in an example of negative feedback.

So-called motor proteins such as actin, dynein, and kinesin are involved in the transport of atoms and molecules from one location in the cell to another, or the movement of extracellular appendages such as flagella and cilia. Video animations of these proteins have gained a lot of attention for their display of what comically appears to be a little robot walking clumsily along a strand of microtubule on two legs and dragging behind it a comparatively enormous "cargo" molecule. The mechanism of this movement has been described in traditional methods of molecular biology require destroying the cell and the natural environment in which proteins and other biological molecules operate; they provide static snapshots of the living cell and its components. Newer technologies provide alternative methods for studying proteins in a more natural state in living cells.

Protein Machines

Since the nineteenth century, cells have been described metaphorically as "chemical factories," and so it is not surprising that their components have been referred to as "gears" or "machine-work," or that microbiologists and biochemists talk about enzymes as part of the cell's "equipment." (At the same time, however, it is worth noting how prevalent agency talk is in chemistry and biochemistry. For instance, the ubiquitous term *chemical reagent* used to denote substances that produce chemical reactions is derived from Latin *agere* – meaning to do or act.) Many of the cell's organelles have been described as machines of various sorts because of the highly specific functions they fulfill, such as the ribosome that synthesizes polypeptides from messenger RNA instructions or the mitochondrion, which is commonly known as the cell's "power plant" or "energy factory" because it is responsible for creating adenosine triphosphate (ATP), the high-energy molecular "currency" that largely drives cellular activity.

Even though proteins are large molecules, they still display the unusual properties that distinguish the microscopic world from the more familiar world of macroscopic objects. For instance, proteins exist in a state somewhere between solid and liquid, which allows them to undergo conformational changes. The solid-like parts of a protein can swing shut or open around hinge-like bonds or rotate around joint-like bonds. The "protein machine" hypothesis emphasizes that "the dynamics of conformational transitions is represented by a quasi-continuous motion along a few 'mechanical' coordinates, e.g. angles describing mutual orientation of approximately rigid fragments of secondary structure or larger structural elements." It trades on the positive analogy between proteins that can change shape by rotating and swinging around particular "joints" and "hinges" and a machine, defined as a "structure which displays high mobility on certain directions and rigidity in others."

In 1998 Bruce Alberts, a leading biochemist who helped determine how DNA is replicated in the cell through the coordinated activity of a complex of different proteins, published an influential essay in the journal *Cell*, titled "The cell as a collection of protein machines: preparing the next generation of molecular biologists." Alberts began by noting that as a graduate student of biochemistry in the 1960s, the image of the cell was basically a bag of enzymes in which separate proteins and other molecules bounced randomly about in the fluid cytoplasm by chance to meet and interact with one another. But later research revealed that many proteins aggregate to form large assemblies of 10 or more proteins that function collectively to carry out specific chemical work. "Indeed," he wrote, "the entire cell can be viewed as a *factory* that contains an elaborate network of interlocking *assembly lines*, each of which is composed of a set of large *protein machines*" (emphasis added). This permits the chemical reactions going on in the cell to occur in a coordinated fashion rather than haphazardly, in such a way that what is achieved in one step is not undone in the next. Alberts then explained the rationale for using the language of machines to describe this coordinated activity:

> Why do we call the large protein assemblies that underlie cell function protein *machines*? Precisely because, like the machines invented by humans to deal efficiently with the macroscopic world, these protein assemblies contain highly coordinated moving parts. Within each protein assembly, intermolecular collisions are not only restricted to a small set of possibilities, but reaction C depends on reaction B, which in turn depends on reaction A – just as it would in a machine of our common experience.

According to the comparison made, like machines, proteins also perform work, and do so in strict conformity with the laws of physics and chemistry, largely being driven by the transfer of high-energy phosphate groups from ATP or other similar molecules (e.g., guanine triphosphate (GTP)). This is contrary to certain earlier notions of vitalism that insisted that life operates according to special laws and forces unaccountable to natural science. But Alberts' chief point was to emphasize how in these assemblies several proteins interact in a coordinated fashion, just as the separate parts of a machine do. And like the parts of a machine mechanism, the proteins are capable of moving relative to one another by changing their three-dimensional shapes without disassembling as a whole mechanism; this allows the substrates on which they act to come into contact with one another, and to do so in a particular order or sequence.

In his earlier work on the DNA replication apparatus of the T4 bacteriophage (a viral pathogen of *E. coli*), Alberts described the complex of proteins involved as a "true replication machine" that moves along the DNA strand base by base, stitching together a new complementary sequence "like a tiny sewing machine" (Figure 4.2). In the T4 bacteriophage a minimum of seven separate proteins work together to replicate the viral DNA: one unwinds the helical DNA molecule into a straight double-strand, followed by a helicase enzyme that cleaves the hydrogen bonds between the base pairs to create two single strands; three more work together as a sliding clamp to keep a DNA polymerase enzyme attached to the DNA strand, while another binds in series to the single strands to keep them from bonding with themselves and forming knots, which thereby provides a straight section for the DNA polymerase enzyme to read with greater efficiency and speed while it synthesizes

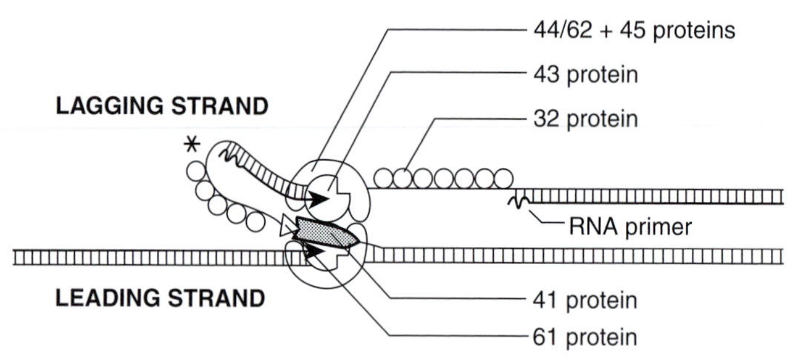

A schematic view of the seven proteins of the T4 DNA replication machine as they are thought to exist in an actual replication fork. The two-dimensional replication fork has been converted into the structure shown by folding the DNA on the lagging strand in such a way as to bring the DNA polymerase molecule on the lagging strand in close opposition to the DNA polymerase molecule on the leading strand. The lagging-strand DNA polymerase molecule is thereby held to the other replication proteins, allowing it to be retained for many successive cycles of Okazaki fragment synthesis.

Figure 4.2 The T4 replication machine (reproduced with permission from Alberts, B. (1984). The DNA enzymology of protein machines. *Cold Spring Harbor Symposium on Quantitative Biology* 49: 1–12, fig.2).

a new complementary strand of DNA from the single-stranded template. (This whole operation takes place on both strands of the cleaved DNA, though it occurs slightly differently on each.) And this complicated arrangement is found in a virus! – which is technically not even considered a living organism since it must infect a live bacterial cell and hijack its energy production system, ribosomes, and precursor molecules to make complete copies of itself. Similar – though more complicated – replication machinery has been found in eukaryotes, including humans. "In retrospect," he wrote, "it seems obvious that complex living systems could not exist without the 'high technology' of multicomponent protein machines."

Alberts' message in his 1998 piece was that students of molecular biology should receive better training in physics and chemistry so that they can exploit the engineering perspective that will allow them to identify "molecular machines" within the cell and explain how they operate according to physical principles. Subsequent research has identified a wealth of multi-component "machines" operating in the cell, such as ATP synthase (which combines a proton "pump" with two separate molecular "motors" to generate ATP, the chief energy "currency" used to make things go in the cell), the spliceosome (which splices together RNA transcripts from DNA exons into messenger RNA), the proteasome (a cylindrical multi-protein structure in which misfolded and damaged proteins are degraded and recycled), and the apoptosome (known as the "seven-spoked death machine" that recruits and activates the caspase enzymes that execute programmed cell death), to name only a few.

Scientists frequently talk more broadly of "molecular machines," a phrase that includes active structures of lesser complexity than the multi-protein complexes discussed above, such as the "motor protein" kinesin and the membrane-bound bacteriorhodopsin protein that is commonly described as a "proton pump." But the rationale for describing these comparatively simpler molecular systems as machines is the same. David Goodsell, a structural biologist, scientific illustrator, and the author of the very popular book *The Machinery of Life*, writes, "The unusual, organic shapes of molecular machines may seem daunting and incomprehensible, but in many ways, molecular machines may be understood in a similar way: as a mechanism where parts fit together, move and interact to perform a given job." He goes on to note that, "Molecular machines, however, have a few fundamental differences from man-made machines, and it is necessary to gain a basic understanding of these differences to appreciate the wonders occurring at the molecular scale."

This is a theme that the philosopher of science Dan Nicholson has recently developed into what he calls the Argument from Scale, which states, to quote him, that "owing to their miniscule size, cells and their macromolecular components are subject to drastically different physical conditions compared with macroscopic physical objects like machines, and that using machine metaphors to explain microscopic phenomena is consequently more likely to obscure and deceive than it is to elucidate and enlighten."

For one, the material from which molecular machines are composed is atoms, not lengths of metal, plastic hosing, or rubber tires. This places unique constraints on the types of devices that can be built, because the atoms from which cells and proteins etc. are made (mostly carbon, oxygen, nitrogen, sulphur, phosphorous, and hydrogen) will only bond with one another in specific ways determined by their chemistry. This also means that the behaviour of molecular machines is far more governed by chemistry than by the macro-scale physics of regular machinery, and the physical forces that do affect them are quite different as well. Because they have so little mass, gravity is negligible, while the constant bombardment due to the thermal (Brownian) motion of the water molecules that surround them batters them to and fro. And of course, most obviously, molecular machines are not designed and built by humans.

Assessing the Machine Metaphor

So how seriously should we take this language of machines? Is it all "just metaphor"? A piece published in the journal *Nature* in 1997 (titled "Real engines of creation") declared that "Some enzyme complexes function literally as machines, and come equipped with springs, levers, and even rotary joints." Evidently some scientists take the language literally. This is not entirely unreasonable, *if* one is willing allow that a machine is anything capable of performing work (and in strictly physical terms, work is simply the application of force to move an object some distance). But it should come as no surprise that there are differing opinions within the scientific community and beyond whether to accept this proposed broadening of the meaning of the term "machine." It needs again to be stated that the machine conception of proteins and molecules is largely based on studies of structure (as revealed by techniques like X-ray crystallography) and of the specificity of binding between enzymes and substrates (which are based on averaging over populations of many molecules studied in controlled *in vitro* conditions). The result is a picture of a generalized and static form (rather than of the dynamic behaviour of individual proteins in their natural environments) that is especially conducive to comparison to rigid machines with fairly permanent parts.

However, as Dan Nicholson emphasizes, more recent developments in technology such as nuclear magnetic resonance (NMR) spectroscopy, *in vivo* microscopy with fluorescent tagging of molecules, and computer simulation of protein dynamics allow researchers to investigate individual molecules in real time and in more natural states, and is significantly changing the picture of proteins. These approaches undermine the idea that proteins typically assume one native structure, suggesting rather that they can fluctuate among a number of compatible conformations, and are strongly influenced by their interactions with other proteins and molecules. Studies of single molecules in real time reveal that proteins are far more plastic, dynamic, and stochastic (i.e., non-deterministic in behaviour) than traditional methods of investigation had led scientists to believe. In fact, a newly described class of "intrinsically disordered proteins" appear to have no permanent native structure at all, and yet seem to serve important functions nonetheless. The lesson here is that the image of proteins and the metaphors used to describe them is quite sensitive to the methods and techniques used to study them. And this is a further reminder that we should regard metaphors in the life sciences as always and only partial perspectives on selective aspects of reality.

Greater attention to the effects of the random molecular motion that tends to dominate at the level of proteins also undermines the picture of proteins as rigid machines with sturdy moving parts. The physicist Peter M. Hoffmann refers to a "molecular storm" to capture the crowded and continual collision of atoms and molecules (much of it water) that makes up the internal environment of the cell. (If you want to try to envision what the cell interior is like at the molecular level, don't think of a pool table in an air-conditioned room with a few balls travelling long distances before striking one another, think of a thick seafood chowder at a simmer on the stove top.) If a similar tumult of stormy weather were scaled up to the size of our regular experience, our machines would be ripped apart (not to mention our bodies as well). To survive in this maelstrom of molecular motion and collisions, proteins and other cellular components have been naturally selected to "bend rather than break," and to exploit the kinetic energy in constructive ways. But even this language obscures the peculiar nature of proteins as "dense-liquids" or "melted solids," not just flexible metals or plastics. The significance of random stochastic effects at the atomic and quantum levels also means that proteins behave quite differently from the macroscopic machines of our everyday experience, where such quantum and Brownian molecular effects are averaged out over the great number of atoms involved, and are therefore unnoticeable to us.

Are Proteins (and Cells) Intelligently Designed?

Another important difference between proteins and regular literal machines is the rather obvious one that molecular machines are neither designed nor built by humans. They are in fact self-assembling, guided by the laws of chemistry and physics, in tandem with the sequence of amino acids of their primary structure that favours certain bonds and folding over others. This level of organization is the result of natural selection for those sequences capable of producing functional proteins that convey advantage to the cells and organisms in which they occur. So while a gene may encode the amino acid sequence of a protein (the primary structure), the higher-order properties (secondary, tertiary, and quaternary) concerning how it folds into a globular shape (or not) and aggregates into larger multi-protein complexes (or not) are not determined by any genetic blueprint or program, but are the combined result of the chemical bonds that form between the amino acid residues (its primary structure) and the context of the internal cellular and broader external environment in which these chemicals interact.

This points to one of the key reasons the biologist and philosopher Massimo Pigliucci and the philosopher of science Marten Boudry are so critical of the use of machine and engineering metaphors in biology. For when we describe proteins as machines or cells and organelles as factories, we encourage the inference that therefore they must have been designed by some intelligent agent; and if not by humans, obviously, then by some other intelligent being. This has been a key argument deployed by advocates of intelligent design, a modern version of the creationist argument for the existence of God famously articulated by the Bishop William Paley at the turn of the nineteenth century. But whereas Paley had to draw the comparison himself between organic structures like the eye and a pocket watch or other human artifact, intelligent design theorists use the scientists' own language to help make their case. If the cell is a collection of protein machines, then some intelligent agent had to design and create these machines, they insist, and what or who could that be but the supernatural creator of the traditional monotheistic religions of Judaism, Christianity, or Islam?

Clearly this argument takes the machine and engineering metaphors used in cell and molecular biology quite literally. We have seen that at least some scientists are also willing to interpret this language literally, on the proposal that we define a machine as any arrangement of parts capable of performing work and exclude any mention of intelligent origins, either human or superhuman. While this proposal may undermine the creationist argument, it does not refute it entirely. For, even if we agree that we need not suppose that all machines are intelligently designed, one is still free to inquire whether these particular machines (those found in the cell) are not the products of intelligent forethought and purpose. In part for this reason, critics of intelligent design recommend against the use of machine and engineering metaphors in biology.

But biologists and philosophers like Pigliucci, Boudry, Richard Lewontin, Johannes Jaeger, John Dupré, and Daniel Nicholson are also concerned that these metaphors are not only misleading for non-scientists looking to interpret the results of science, but for the scientists themselves. Here again the emphasis is on what the philosopher of science Mary Hesse called the negative analogies in the organism–machine comparison. The practice of describing organisms, cells, and their components (organelles, proteins, genomes) in the language of machines obscures just how non-machine like, and how non-intelligently, non-optimally, and just plain different organisms are from a "design" perspective. And yet these metaphors may still be regarded as heuristically valuable insofar as they act like hypotheses that can be put to the test, and should they be found inadequate, scientists will have at least learned important ways in which cells and proteins are unlike machines.

In their attempts to educate the public about how evolution by natural selection proceeds, biologists have invoked their own instructive metaphors. Richard Dawkins, for instance, has described natural selection as "the Blind Watchmaker" that assembles living cells and organisms from available materials with no forethought or plan for the ultimate outcome; and François Jacob has compared evolution to a "tinkerer" who bangs together arrangements of available materials that are simply good enough for the immediate job at hand, not optimally or intelligently designed as an engineer would do by gathering together only the best materials and arranging them in the most efficient way possible. Cells and organisms are complex systems, but they are frequently a mess from a rational design perspective. Signal pathways, for instance, by which the genes stored away in the nucleus are switched on or off in response to extracellular signals, have been compared to Rube Goldberg machines, the overly complicated cartoon creations that eventually manage to accomplish a simple task such as switching on a light.

But as Pigliucci points out, even this is not an entirely apt comparison, because despite the similarity that both systems are complicated, the human-engineered Goldberg system is easily derailed if just one component fails to operate as it should (such systems are said to be "brittle"), whereas cellular systems have built-in redundancies (they are said to be "robust"), so that if one component molecule or protein fails to perform a vital task, another fills in to ensure it is carried out. This feature is not a reflection of intelligent design, but is easily explained as the result of natural selection; since cells or organisms with brittle systems would be at a disadvantage to those that have managed to put together, albeit clumsily, more robust systems and these would be passed on to more descendants in future generations. But the continued use of machine and engineering language to describe and to understand the physiology and anatomy of organisms and their cells, Pigliucci and others warn, may prevent scientists from properly understanding instances where "the object of study becomes so remote from everyday experience that analogies begin to do more harm than good" and send them on a "wild goose chase."

Messengers and Team Players

So far we have seen how proteins are frequently described and understood "through the lens" of machine metaphors. But as explained in Chapter 2, the background metaphor of human agents and agency has also been very influential in the life sciences, and this applies to the study of proteins as well. In addition to functioning like machines, many proteins play important roles in the communication between cells throughout the plant and animal body. This allows the organism to regulate its activities and to respond to its environment as a coherent collective whole, not just as a bunch of disconnected cells. In animals this can be achieved through the nervous system, which relays electrical and chemical signals from one organ to another, but it also occurs through the endocrine system by means of hormones passing through the bloodstream, or even more directly between adjoining cells in the same tissue.

The development of a complex multicellular animal, like us, with differentiated and specialized tissues and organs, proceeds from a single fertilized egg cell, through multiple cell divisions. In its very early stages, therefore, the embryo is a clump of genetically identical cells, and not surprisingly, these early cells look identical. But gradually these cells begin to express specific genes for growth factors and transcription factors, which they will send to their neighbours. These signals will have the effect of switching on or off specific genes in the right cells at the right times so that populations of cells in different parts of the embryo begin to develop down different developmental pathways. (In animals, the mother inserts a non-uniform gradient of transcription factors into the egg, so that even before the first cell division occurs there will be subtle differences between the two daughter cells and the subsequent fates of their cell lineages. The first few rounds of cell division are thus under the direction of these so-called maternal-effect genes.) The continued growth, repair, and normal function of adult organisms relies on continuous intercellular communication.

Many proteins function as transcription and growth factor signals in cell-to-cell communication. As mentioned earlier, an external "signal" binds to a "receptor" on the cell surface (although some are able to slip right through to the cell interior), is "transduced" across the cell membrane, after which it "triggers" an entire "cascade" of second "messengers," kinase "switches," and other molecular complexes that comprise a signalling "pathway" leading into the nucleus. Because of a large degree of interaction among these pathways, referred to as "cross-talk," biologists increasingly talk of signalling "networks." Note all the metaphors here! And the different source domains from which they draw. Proteins play roles in all these events, from the receptors to the second messengers, the switches that activate other proteins in the pathway, and so on. Because the proteins in these signal pathways often aggregate to form multi-protein complexes they are frequently said to "recruit" one another to perform a specific task, after which they may disassociate to join up with yet another team to perform some other function.

What does "recruitment" really mean here? Are proteins calling out to one another, saying "Come on everyone, it's time to go to work to achieve objective X?" No. In fact, scientists do not typically speak of proteins signalling to or communicating with one another; proteins are said to *interact* with one another. Proteins do, however, "associate" with one another on the basis of what are called peptide signals (i.e., specific amino acid sequences) that recognize or bind to matching receptor sites on other proteins, DNA sequences, membrane lipids, or other sites in the cell. So when two or more proteins interact in the sense that one recruits the other(s), what is meant is that they form chemical bonds with one another (at these sequence-specific sites), creating a multi-protein complex that carries out some specific function. This is another example of what initially looks like teleological or purpose-driven activity in the cell, which explains the employment of agential language to describe it. It can, however, be accounted for as another instance of *teleonomy* – that is, the atoms of which these molecules are composed form chemical bonds through the sharing of electrons in their outer shells, and the interactions among molecules that have increased survival and replication of the cells in which they occur have been naturally selected (metaphors involving the notion of natural selection will be discussed in Chapter 6).

It is helpful for us to describe these interactions in purposive, social, and agential terms – since we are goal-driven, highly social agents – but it is important to recognize that these are metaphors. Kostas Kampourakis puts it well when he writes:

> Cell biologists refer to "signaling" proteins and to "receptor" proteins to describe their chemical interactions that bring about changes in the status of the cell. Whereas the two proteins simply interact chemically, they are described as "signal" and "receptor" because the outcome of this interaction is a change, or a series of changes, within the cell. Therefore, scientists describe the protein that makes the difference in bringing about these changes as a "signaling" protein, *as if* this protein is transferring a signal for initiating these changes. They also describe the other protein as a receptor, *as if* it receives the signal transferred by the other protein ... [However] one should keep in mind that signaling proteins are not really active agents in intercellular communication, do not have any intentions, and their interactions and subsequent changes depend on several other molecules around them.

Another example where proteins are described as if they are social agents is the metaphor of "cooperative" binding. Molecular biologists speak of "cooperativity" as the property whereby the binding of a molecule to one site on a protein increases the affinity or chances of a similar molecule binding to a site at another location on the same protein. This is an example of allostery (defined by François Jacob in the 1960s as an "interaction at a distance") that occurs when the binding of a ligand at one site causes a conformational change in the protein favourable to the binding of another ligand at a distant site. But cooperativity can be used more broadly to describe the separate activity of a number of atoms or molecules that collectively bring about some result that otherwise would not occur or would be far less likely to occur. The cell is a crowded and busy place, and what is described metaphorically as the cooperative behaviour of its components is key to its success. Scientists regularly describe the behaviour or activity of molecules and macromolecular complexes using agent metaphors in order to highlight their contribution to some function of interest. But in a strictly literal sense the activity ultimately comes down to chemistry and natural selection (i.e., the differential survival and replication of adaptive systems of chemical reactions).

As a general rule, it seems that agent metaphors are used in molecular biology to describe the behaviour of a protein or molecule, while machine and engineering metaphors are used to describe how they manage to achieve that activity, or what is commonly called the *mechanism* involved. The different background metaphors then appear to fulfill slightly different functions in the context of scientific investigation. As a recent review paper puts it, the traditional approach has been to focus on all the various molecular actors in a system and through experimental interruption discern their functional roles in that system. The authors call this "Shakespearean biology," which they contrast with their preferred "Newtonian biology," wherein scientists articulate mathematically precise descriptions of the behaviour and interactions of the actors so as to make quantitative predictions.

In either case, whether they draw on our experience with human-made machines or human behaviour, it is important to remember that the metaphors used to describe proteins will always offer partial and selective perspectives on the reality in question. In that respect, the metaphors scientists use to talk about and to understand their subjects of study should be regarded as provisional hypotheses that may need to be revised or even rejected if they fail to be adequate to the empirical and experimental data or otherwise useful for making sense of it. On the other hand, because metaphors also function as convenient labels that facilitate communication among scientists and with non-scientists, there is always the chance that metaphors that have been shown to be empirically inadequate will remain in currency like a bum coin or counterfeit note.

5 Cells

Factories, Computers, and Social Organisms

We have talked about genes and proteins and the various metaphors used to describe what they are, and how they function and interact with one another in the context of the cellular environment. Now it is time to look at the cell itself – the fundamental unit of life, the minimal system to which the property of being alive can be ascribed. While DNA and proteins are complex molecules with the capacity for chemical activity that is vital for life, neither of them can be said to be living. Nor can any other component found beneath the level of the cell as a whole: not amino acids (the components of protein), not polysaccharides (the large chains of sugars of which starch, cellulose, and chitin are composed), not water (which makes up more than 70 percent of the cell by content), not lipids (the fatty molecules that make up the membranes that surround the cell and many of its internal organelles), nor the organelles that carry out specific tasks within the cell. (A possible exception might be made for mitochondria and chloroplasts, organelles that evidence suggests were once themselves independent living cells that somehow came to live symbiotically within another larger cell.) Only the whole system taken together and properly organized is capable of all the properties that collectively define a living being: growth, metabolism, reproduction, motility, ability to maintain homeostasis (a stable internal state conducive to life), adaptive responsiveness to the environment, and ability to evolve. According to the cell theory, widely regarded as one of the greatest achievements in the history of science, you and I are alive and able to do the amazing things we can do because our cells are alive and able to do the amazing things they can do.

Cells are sometimes called the "atoms" of biology because they are the "building blocks" from which all forms of life are made. But this does not mean cells are simple in composition or behaviour. On the contrary, cells are complex systems, and so it will come as no surprise that a rich variety of metaphors have been used to describe them. The cell is often called a "factory" engaged in the production of proteins and other biomolecules or described in the terms of an electronic computer that is controlled by a genetic "program" whose "instructions" are "encoded" in DNA, and conveyed along "signalling pathways" and genetic "circuits." But like proteins, the cell's behaviour is described in both machine and agential language. For instance, cells are said to make "decisions" about what sort of cell to become or whether or not to engage in activities like cell "suicide," a genetically regulated process alternatively known as "programmed cell death." Cells have been regarded as "elementary organisms" joined together to form specialized tissues and organs, whose "division of physiological labour" makes possible the higher plants and animals; and from this perspective, each of us is a "society of cells." Our cells are able to achieve this great level of specialization and integration by "communicating" with one another through various mechanical, chemical, and electrical "signals." The development of multicellular animals like us begins with the division of a fertilized egg cell into a clump of "stem" cells, from which "lineages" of "daughter" cells descend various "paths" in a developmental "landscape" to arrive at their ultimate "fates" as specialized tissue cell types.

But Why Are They Called "Cells"?

Given that cells are engaged in such a wide assortment of behaviours, it's no wonder so many metaphors have been used to describe and to make sense of them. But the original notion of the cell was itself quite simple. It is also instructive of an important lesson about the role of metaphor in science. Just ask yourself, "Why do we call these fundamental units of life, these elementary organisms, these complicated little computer-like factories, why do we call them *cells*?" Many people, I suspect, will not have realized, unless they are familiar with the history of cell biology, that even the term "cell" is a metaphor. Or rather that it began as a metaphor, when in 1665 the English naturalist Robert Hooke, in one of the first written accounts of observations made with the newly developed compound microscope, described the tiny little boxes he saw in the tissue of dead cork plant as "cells," because they reminded him of the hexagonal cells of a bee's honeycomb. (Hooke does not in fact mention the cells in which monks live as many textbooks and other accounts claim. But the chambers in honeycomb are likely called cells because they reminded people of the little rooms in which monks and prisoners live.) Hooke also described the hollow spaces as "bladders," "boxes," "bubbles," "caverns," "chambers," and "pores." What caught his attention was the appearance of a series of little spaces distinctly separated from one another by rigid walls.

Hooke did not propose that all plants or all animals are composed of such cells, for the very good reason that he could only see such structures in a few organic specimens like cork and the pithy material found in the shaft of a bird feather. For the next 150 years or so, naturalists continued to use a wealth of different terms to describe the spaces, sometimes empty, sometimes filled with sap or juice, discernible in many plant tissues, including "cavity," "globule," and "vesicle," in addition to those already mentioned by Hooke.

So, it seems, "cell theory" was no inevitable development, at least so far as terminology is concerned. We might today be speaking of the box or chamber theory, were it not for the contingent fact that two German scientists, Matthias Schleiden (a botanist) and Theodor Schwann (an animal physiologist), used the German equivalent *Zelle* in what became very influential essays they published on plant and animal development in the early nineteenth century. In 1838, Schleiden declared that all plants are composed of and by microscopic *Zellen* ("cells" in the English translation of 1848), which he declared to be living organisms and individuals in their own right, but also capable of leading a second kind of communal and dependent life as parts of a larger multicellular plant individual. The plant, from this perspective, was a kind of composite "super-organism." Robert Brown's description of the nucleus in 1833, a small body visible in plant cells, helped Schleiden to identify the presence of cells throughout the diverse structures and forms of plant tissue. Schwann then extended this idea in 1839 to animals, declaring that all living things are composed of and by cells. As evidence for this very broad and bold claim, Schwann pointed to various stages of embryonal and fetal development in different animals that displayed a honeycomb-like partitioning similar to that first highlighted by Hooke. Schwann could also point to more mature structural units in animal anatomy (such as epithelial cells lining organs and the outer skin) that exhibited the clearly defined enveloping wall to which the term "cell" so pointedly drew attention. In that regard, the metaphor provided scientists with a valuable search image to guide their observations through the microscope. But in order to convince himself and others that the concept of the cell was universally applicable to the entire animal, even when so many tissues and organs showed no sign of clearly defined walls separating one unit from another, Schwann construed the cell as a developmental unit, an elementary organism whose merging together and transformation into continuous sheets of multi-nucleated living matter could account for the gradual development of all the different tissues, organs, and other structures in the adult animal body, regardless of whether they displayed a cellular appearance. Henceforth, the cell theory attracted much attention, criticism, and ultimately many adherents.

"The Cell" Is Dead – Long Live the Cell!

But the term "cell" itself was the target of much criticism throughout the nineteenth century and into the next, precisely because so many of the living things that were supposed to fall under that descriptive term lacked any indication of a solid wall. Many single-celled organisms, protozoans like amoebae, and even white blood cells in our own bodies, are constantly changing shape and oozing about, something they could hardly accomplish were they boxed in by a solid wall as in Hooke's original cork cells. And many other types of animal tissue, for instance, muscle and nerve fibres, do not appear to be composed of many individual cells but a continuous network of protoplasm, the term for the "physical stuff of life" that became popular from the 1860s on. Various alternative terms were suggested to replace "cell," from *bioblast*, *energid*, or simply *clump of protoplasm*. However, none were able to dislodge "cell" from the minds and language of biologists.

After all this time it is no longer necessary to worry about the negative analogies suggested by the term, because what was once a metaphor is no longer. "Cell" is now a dead metaphor, which is to say that when scientists and students use the term they do so quite literally. No one today, when asked to think of a cell in the context of biology, thinks of prisons or monk's cells. The term now just literally means "the fundamental unit of life." So here is an important lesson: The distinction between metaphor and literal terminology is not a hard and fast one; one can transition to the other, even in science. Repeated use and consolidation of definition is one means for this to happen, but it is not the only way. Rather than changing how we think about the metaphor, in some cases the metaphor actually ends up quite literally changing the thing to which it is applied. This is what happened with the metaphor that "cells are chemical factories."

Cells Are Chemical Laboratories or Factories

Around the middle of the nineteenth century, some physiologists interested in understanding internal cell function began comparing cells to a chemical laboratory. It was clear that there is lots of chemistry going on inside the cell, and drawing analogies with a laboratory full of chemicals and flasks helped shed light on how plant cells manage to create complicated sugar molecules from simpler materials like water and atmospheric gas (CO_2), or how animals are able to digest and transform food materials obtained through the consumption of other plants and animals into the cells, tissues, organs, and bones, etc. of their own bodies. Assuming that the cell operated on the same physical and chemical principles as does the chemist in his or her laboratory was an important step toward overcoming vitalism, the thesis that life operates according to its own special, and perhaps non-physical or non-materialistic, principles that could only be explained by appeal to a supernatural creator. Schwann, himself an animal physiologist, introduced the term "metabolism" to describe the chemical processes of digestion (katabolism) and the synthesis of new compounds from the recycled products and energy attained from digestion (anabolism). According to the cell theory, all this chemical activity is carried out by and in cells.

It was not long before physiologists working in the industrial settings of nineteenth-century European cities began describing the cell as not just a chemical laboratory, but an industrial factory. Rudolf Virchow, for instance, the professor of pathological anatomy and physiology in Berlin who would do much to spread the reception of cell theory into medicine and the other life sciences, said in 1858 that "starch is transformed into sugar in the plant and animal just as it is in a factory." Similarly, the influential French physiologist Claude Bernard, in describing the cellular structure and function of animal organs, in 1885 wrote, "In the living body these organs are like the factories or the industrial establishments in an advanced society which provide the various members of this society with the means of clothing, heating, feeding, and lighting themselves."

What were the positive analogies to suggest such a comparison? Cells import raw materials and use them to create new products, which are eventually exported outside the cell and shipped throughout the plant or animal body through the sap or blood. But more importantly, the laboratory and factory metaphors suggested some very interesting neutral analogies to pursue experimentally. From around the turn of the twentieth century, biochemists had begun investigating the chemical activity of various molecules called enzymes that could be extracted from cells. In 1897, Eduard Buchner showed that fermentation of sugar into alcohol could occur without the presence of whole live yeast cells by extracting from them a juice containing the active agent, which he called zymase. Biochemists began identifying other enzymes capable of important metabolic function by grinding up various types of cells, removing the fluid contents, and investigating the chemical activity of the ingredients *in vitro* (in glass test tubes). It became evident, though, that while biochemists could in this way replicate in their test tubes the metabolic reactions of living cells, they could not replicate them with the efficiency or productivity achievable by the whole cell. Biochemists began to realize that these "grind and find" techniques, as they came to be called, were limited in their ability to reveal how living cells operate, not only because they required killing the cell, but because they treated it as though it were merely a "bag of chemicals" in which enzymes and their substrates float about to meet haphazardly in no particular spatial or temporal order.

But if the cell really was comparable to a laboratory or a factory, then surely it ought to be properly organized so that the materials and tools needed for the various operations are spatially situated near one another, like the chemist's bench or the factory floor; and in that way the construction of products can proceed in an efficient and orderly sequence, moving perhaps even from one workstation to the next along a sort of assembly line. The laboratory and factory metaphors, therefore, suggested that, far from being a bag of enzymes loosely sloshing about, the cell is a highly ordered system with an internal architecture designed to facilitate the speedy and efficient execution of vital chemical reactions. Protein synthesis, we now say, is carried out at the site of the ribosomes, themselves frequently referred to as factories for the production of proteins, which are situated on the rough endoplasmic reticulum adjacent to the nucleus. After translation from the messenger RNA transcript, the polypeptide is carried through the endoplasmic reticulum as if by conveyor belt for post-translational modifications (proper folding, addition of functional chemical groups or signal peptides that act like shipping labels indicating where the protein is to be transported, etc.). The protein may then be moved along to the Golgi apparatus, where it is "packaged" into a lipid vesicle for transport perhaps to the cell membrane and export out of the cell.

Today, it is a common school exercise to have students learn about how the cell is like a factory or industrial plant that manufactures proteins, and to identify each of the main cellular components with its analog in the factory system. This analogy is frequently portrayed in cartoon form as a visual metaphor, as seen in Figure 5.1. (Note the combination of factory and agent metaphors: the proteins are depicted as little beings running organelle-machinery. Recall that enzymes and other active molecules are regularly called chemical *agents*.)

Genetic Engineering Turns Cells into Literal Factories

Humans have for millennia harvested the natural ability of organisms to synthesize various biochemical products, none more popular perhaps than the lowly yeast's ability to turn sugars into alcohol such as beer and wine. On a more sober note, the discovery of hormones and their role in the normal function of the human body spurred attempts to harness these molecules for the treatment of disease and poor health. Insulin was identified in 1921 by Frederick Banting and Charles Best as the chemical ingredient produced in the pancreas, responsible for regulating levels of sugar in the blood, a deficiency of which leads to diabetes. For several decades, insulin was harvested from cattle and pigs for use in the treatment of human diabetes, but in the late 1970s the development of recombinant DNA technology allowed researchers to splice the gene for human insulin into yeast and bacteria so that these cells produced the hormone. Today, recombinant human insulin is produced on an industrial scale in large vats of genetically engineered bacteria and yeast. This was the first big success story of the biotech industry, and it was made possible by taking the natural ability of cells to *act like little*

Figure 5.1 The cell visualized as a factory (Nigel Sussman, *Science* 366(6467) 15 Nov. 2019 cover; reprinted with permission from AAAS).

factories and modifying it so they churn out a different product of greater value to us humans. In other words, the cell went from being a *metaphorical* factory to a *literal* factory with the removal of a key negative analogy, that factories are human artifacts and not natural systems. By engineering these cells to express human insulin, they became at least in part human artifacts. Today there is a journal, *Microbial Cell Factories*, published by Springer Nature, devoted to innovation in biotechnological applications of genetically engineered cells; and companies like Thermo Fisher Scientific market "Cell Factory Systems" used in the production of vaccines and other "biologics designed to improve and save lives." What were once just metaphorical cell factories, are now quite literally saving human lives.

This brings us to the other means by which a metaphor can become a literal term in science. Earlier we saw that a metaphorical term like "cell" can become literal when the negative analogies and misleading connotations associated with the original metaphor fade from people's attention, leaving us with a "dead" metaphor. What happens in such a case is that the metaphor changes the way we think about the object in question. We originally look at the biological units and see them as little chambers or vessels in which the powers of life are contained, and gradually as we learn more about the object the meaning of the term changes from metaphorical to literal. Talk of the cell just literally refers to the fundamental unit of life. But in the case of the cell factory metaphor, what has happened is that the metaphor has not only changed *how we think about* the object to which it is applied, but has also eventually changed its very *nature or reality*. For what were once only metaphorically regarded as being like factories, have now literally become factories.

Metaphors are powerful tools indeed. We will return to this theme in Chapter 8 when we discuss the metaphors of reprogramming cells and rewiring cell circuits in the context of biomedicine.

The Society of Cells and the Cell-State

Lest one get the impression that cells have forever been regarded as little machines or other technologies, we now turn to another set of metaphors that have been very influential in their own way. Recall that two of the important founders of the cell theory (Schleiden and Schwann) described cells as little organisms. From this perspective, it would appear, then, that each multicellular plant or animal is a composite entity, more similar to a society of individual cells than one coherent and indivisible organism in and of itself. This was a popular talking point in the nineteenth century among biologists and philosophers. Rudolf Virchow described the human body in 1858 as a "Cell-State" or "society of cells, a tiny well-ordered state, with all of the accessories – high officials and underlings, servants and masters, the great and the small"; and Herbert Spencer a decade later referred to the body as a "commonwealth of monads" (a popular term for unicellular protozoans and protophytes). Ernst Haeckel, one of Virchow's students and known in his time as "the German Darwin," wrote extensively on this theme, even distinguishing between plants as "cell-republics" and animals as "cell-monarchies," because in the latter the activity of the cell-citizens is under the centralized rule of the brain and its neuron cells.

In the context of nineteenth-century German politics, this talk of the body as a unified state of individual cells working together harmoniously toward the greater good had real appeal, especially for progressives like Virchow who, in the years leading up to the unification of the different German states and principalities into a unified country in 1871, campaigned for a republican constitution over a more traditionally Prussian and repressively authoritarian model. But in addition to emphasizing the relative autonomy and worth of each individual citizen in the state, for Virchow the cell theory also provided a fresh approach to understanding disease and pathology in the human body. In contrast to traditional humoral accounts that attributed illness to imbalances in the volume and influence of bodily fluids, Virchow became a champion of cellular pathology, according to which the symptoms of disease in the body are traced to dysfunction at the level of the cell. For biologists like Haeckel, who was an early convert to the Darwinian theory of evolution, cell theory and the thesis that the animal and plant body is a state or society of cells motivated an entire research program of explaining how these higher-order multicellular individuals had evolved from independent single-celled ancestors like the amoeba and other protozoa. Haeckel's attempts to gather evidence of this humble ancestry through inspection of the developmental stages of human embryogenesis drew fervent criticism from the politically and religiously conservative, but encouraged many young scientists to study developmental biology and to construct phylogenetic trees that would reconstruct the evolutionary history of life on earth.

While the cell-state or society of cells metaphor has lost much of its political resonance, it continues to motivate a novel perspective on questions in developmental biology, the evolution of multicellularity, and the origins of cancer by focusing attention on the interactions between individual cells and the transitions that occur between individual- and population-level behaviour. This is an approach strikingly called "cell sociology." It serves to emphasize that cells behave quite differently when they associate and act together, and are capable of novel and impressive feats of which they are incapable in isolation. These are known as group or population effects. For instance, tissue cells can be removed from the human body and survive in a dish in the lab if provided with the right temperature and culture medium. (You may have heard of the HeLa cell line derived in the 1950s from the ovarian tumour of a young Black woman – without her knowledge or permission – which, because it had accumulated several cancer-enabling mutations has become "immortal" and continues to divide and grow in labs around the world long after her unfortunate death.) But if one normal tissue cell is wholly isolated from its neighbours, it tends to revert to a more primitive, less differentiated state. And if the culture medium does not provide it with certain signals it would normally receive from its fellow tissue cells (growth factors, survival factors), the cell will shrivel up and die, in an act called "cell suicide." The ability to differentiate as specialized tissue cells – and even to continue to live – are, among multicellular animals like us, group effects achievable only by a "society of cells." We are indeed stronger together.

Cell Suicide and Programmed Cell Death

Embryology, the study of the development of a mature animal from a single fertilized egg cell, revealed the careful and regular timing of significant events in the creation and molding of tissues and organs, all carried out by the division, transformation, and movement of individual cells, often working in concert as groups identifiably similar in terms of morphology and patterns of gene expression. Haeckel's drawings of various stages of embryo development, arranged in rows and columns to allow for comparison between distinct species of vertebrates, helped to make the process of embryogenesis famous. In the early stages of human development (weeks 4–6), pharyngeal arches (otherwise known as gill arches because in fish they develop into the gills) emerge and gradually disappear as they transform into structures of the lower jaw and upper neck; a notably long tail decreases in length until there are typically only several short vertebrae upon birth, known as the coccyx; and around week 8, individually separate toes and fingers are created from what begin as broad paddles, as the cells in between the bones of the digits fade away. All of these processes involve a highly regulated and predictable form of cell death.

The death of cells in developing animals had been observed in the late nineteenth century, but it was not until the mid-twentieth century that it was recognized as a widespread occurrence of fundamental importance in the normal development and maintenance of animal physiology. In the early 1960s, as a doctoral student at Harvard University, Richard Lockshin was working with his supervisor, Carroll Williams, on the developmental changes in the pupae of the silkworm moth. Lockshin observed that the cells of the intersegmental muscles underwent a precisely timed death, which suggested to him a mechanism under genetic control. Influenced by Jacob and Monod's notion of a genetic program he called it "programmed cell death." As he explained many years later:

> Because computers were just beginning to be talked about at the time, programmed cell death seemed to be a particularly modern and colorful way of describing what we saw. It was a metaphor stating what I thought was pretty obvious – if a biological process occurs at a defined location and time, then it must in some fashion be programmed or written into the genetics of the organism – but, as in poetry, metaphors help people see things that they otherwise would not have noticed. Thus a relatively straightforward observation gained some currency.

The metaphor of programmed cell death was taken up shortly after by John W. Saunders Jr., who was studying developmental changes in the limb bud of the chick embryo. Like Lockshin and Williams, Saunders discovered not only that cells in a specific region of the wing limb as it extended from the main body underwent a precisely timed die-off, but also that he could save these cells from their "death sentence" by excising them and transplanting them to another region of the embryo. This worked only if they were transplanted before a specific stage in development, as if a "death clock" counted down the seconds to the crucial moment after which no repeal could be granted. Saunders postulated that the cell's destruction might be due to the release of toxic chemicals from its own lysosomes, the membrane-bound organelles in which viruses, bacteria, or worn-out cell components are degraded by digestive enzymes. The molecular biologist Christian De Duve had dubbed lysosomes "suicide bags" and Carroll Williams had described them as "biological booby traps" because they were lethal to the cell if they released their contents into the cytoplasm. Because the cells Saunders was observing appeared to be actively involved in their own deaths, he described it as "cell suicide." However, given that their fates seemed to be determined by their location in the embryo – and so by the influence of their cell neighbours – Saunders (and later other biologists) also pondered whether it might not be more accurate to consider this a case of "assassination" or "execution" rather than suicide.

The study of active cell death (distinguished from passive death due to injury or disease) became a hot topic in the 1990s with the founding of several journals devoted specifically to its study. In the early 1970s, three pathologists (John Kerr, Andrew Wyllie, and Alastair Currie) published a paper describing the diagnostic morphology or observable features of cells undergoing programmed cell death. Such cells display a characteristic shrinking of the cytoplasm, fragmentation of the nuclear chromatin into regular-sized chunks, bulging of the cell membrane (which they dubbed "blebbing") reminiscent of boiling water, dissolution of the cell into small membrane-bound bits, and eventual consumption by neighbouring cells. They gave this collection of events the name "apoptosis," from the ancient Greek meaning "to fall away," in analogy with the falling of dead leaves from a tree. Unfortunately, as it turned out, the term apoptosis would prove to be very popular and come to be used interchangeably with the expressions programmed cell death and cell suicide (more about this in a moment). Apoptosis and/or programmed cell death are a regular feature of embryogenesis and maintenance of normal and healthy tissue and organ function in animals, plants, and other organisms. In fact, as you are reading this sentence, millions of cells in your intestines, bone marrow, and elsewhere are killing themselves.

As a general rule (there are exceptions), apoptosis or programmed cell death involves the cell expending its own energy to synthesize the enzymes (cysteine-aspartic proteases, or caspases for short) that will cut it up into pieces for consumption and recycling by neighbouring cells. Moreover, in contrast to death by injury, in which the toxic contents of the lysosomes can lead to fatal damage to neighbouring cells, apoptosis is a highly controlled form of self-destruction that has been compared to the controlled demolition of a building by implosion (contrast that with the consequences of an explosion for neighbouring buildings). Because it plays such a vital and constructive role in animal embryogenesis, programmed cell death, under the catchier description of cell suicide, was frequently portrayed as an altruistic act on the part of the cells undergoing it for the benefit of the larger cell society. While this was an effective way of framing the broader social context in which this activity occurs, it ultimately proved to be rather misleading, for it prevented scientists from recognizing that it could appear outside of the social context of multicellular organisms. Microbiologists, for instance, were initially under the impression that programmed cell death would play no role in the life cycle of unicellular organisms, because it seemed counterintuitive that a single-celled organism that is not part of a multicellular organism would kill itself for altruistic purposes. Carefully discriminating programmed cell death from cell suicide, however, allowed this possibility to be more visible. Many bacteria and other microbes as a matter of fact do live as part of larger communities, forming what are known as biofilms that provide mutual protection from predators or attack from the immune system of a host organism in which they might be living (such as in our own guts or mouths). Engaging in programmed cell death in these conditions may be beneficial for the community of cells (with which it may be closely related) and result in the production of various "public goods."

Additionally, evidence suggests that the mechanism for programmed cell death may have originated with bacteria as a kind of "addiction module" encoding both a toxin and its antidote that would destroy any cell that attempted to expel the invasive pathogen. Over time, the genes for this toxin–antitoxin module may have been integrated into the host cell genome, in an example of endosymbiosis. To this point, it is noteworthy that mito-chondria (the "powerhouse" organelles that produce the bulk of ATP energy for the eukaryotic cell) are widely regarded as endosymbiotic descendants of once independent prokaryotes, and they play key roles in the initiation of programmed cell death through the activation of the caspase enzymes that cleave the cell's contents into small bits. Once this toxin–antitoxin module was part of the eukaryote cell line, it may have been co-opted for the advantage it conferred on early colonies of eukaryotic cells negotiating the transition toward being a true multicellular organism. So-called cheater cells that sought to divide for their own reproductive benefit at the expense of the reproductive fitness of the society of cells as a whole could thereby be controlled.

Programmed cell death can be distinguished into two forms: intrinsic, whereby the signal that initiates death comes from within the cell (as a result of internal damage or distress, for instance); and extrinsic, in which the cell receives either a signal ("order") to kill itself from other cells or fails to receive signals to continue living. The extrinsic form has been characterized as assassination or execution. The cell biologist Martin Raff has described it as a form of "social control" exerted by the organism over its constituent cells, the breakdown of which, by genetic mutation to the proteins involved, leads to the selfish and "anti-social" behaviour of cancer cells.

The practise that emerged of using apoptosis, programmed cell death, and cell suicide as synonyms proved unfortunate because they refer to distinct forms of active cell death or the contexts in which it may occur. "Apoptosis" refers to a particular *morphology* of cell death, whereas "programmed cell death" refers to a genetically regulated *mechanism* of cell death. And "cell suicide" is a particularly evocative (some would say provocative) way of referring to the *context* in which either (or both) of the other two phenomena may occur. Not all instances of apoptosis are the result of the initiation and execution of a genetically encoded cell death program, and not all programmed cell death displays the characteristic features of apoptosis. Additionally, the altruistic act of cell suicide may just as reasonably be described as assassination or self-execution under duress, and may have evolved from a behaviour originally more analogous to a hostile terrorist hijacking. Suicide is also a complex and sensitive issue laden with social, cultural, and emotional dimensions that make casual use of the term problematic. Recently, there has been growing awareness within the media that the once common practice of describing someone as having "committed" suicide suggests a voluntary act, which in the age of cyber-bullying and better awareness of mental illness is problematic. Scientists, then, may wish to weigh the unintended social impact of using the language of suicide against its positive contribution to scientific understanding.

The Cell as Computer

Talk of programmed cell death highlights how computer and programming language found its way into cell biology and has strongly informed our understanding of cell function and behaviour. Cells are said to "initiate and execute" their genetically encoded death program upon the reception of so-called death signals, or alternatively, the failure to receive "survival signals" from fellow cells. As we saw in Chapter 4, signals received from outside the cell are said to be "transduced" across the cell membrane, and once inside they can trigger various signalling pathways or genetic circuits that impel the cell to grow, divide, express or repress the synthesis of various proteins, continue to live or to die. And illustrating yet again how contingent current science is on the technological and social developments of its time, there is now talk of a "cell-wide web" of communication centred on the nucleus of myocytes or muscle cells by means of calcium ions (Ca^{2+}) shuttled along an "intranet" system of nanotubes that act like tiny wires extending throughout the sarcoplasmic reticulum of the cytoplasm.

Scientists' ability to identify the various steps and molecular components of these signal pathways – and to manipulate the "logic" of these "circuits" by "switching" them on or off by means of optogenetics or other interventions – has encouraged the discourse of "rewiring" and "reprogramming" cells. Synthetic biology is the express effort to reverse engineer cells as though they were tiny computers running on genetic programs so that they can be reengineered for human purposes, such as turning them into factories for the production of various bioproducts or switching off the faulty program that results in cancer (or alternatively switching on the death program to stop its further growth and spread). These engineering projects will be taken up in Chapter 8, dealing with metaphors in biomedicine.

Stem Cells

Stem cells have become a topic of great significance, mostly because of their promise as a biomedical therapy for treating any number of health conditions. But the term or concept of stem cells is itself rather old, tracing back to the nineteenth century. The German evolutionary biologist Ernst Haeckel coined the term "Stammzelle" in 1868 to denote the original single-celled ancestor of all current life. The metaphor refers to plant growth (trees specifically) as its source domain and projects the associations of an ancestral "family tree" or genealogy to the related target domains of organismal development and species evolution, both of which were covered at the time by the German word *Entwicklung*. The notion of a "family tree" is itself a metaphor based on the analogy of a many-branched tree (representing the many individual descendants) developing from an original stem source of ancestors. Haeckel promoted the thesis that the history of the relationship between species (which he called phylogeny) was imprinted, albeit imperfectly, within the development of individual organisms (ontogeny). Hence the phrase he made famous, that "ontogeny recapitulates phylogeny."

Haeckel invoked the metaphor of a tree stem to help visualize the relation of descent that exists between (1) an original single-celled ancestor or Ur-organism from which all extant life on the planet has descended, and (2) a fertilized egg cell that repeatedly divides to become a multicellular plant or animal with specialized tissues and organs. Darwin of course had earlier used the metaphor of a "tree of life" to illustrate his theory of the community of descent (more about all this in Chapter 6). Haeckel adapted the tree metaphor to capture the thesis that in both the development of new forms of life over the entire history of life (phylogeny) and the development of new organisms (ontogeny), there is a continuous line of descent from an original "stem cell." Today, talk of stem cells is typically confined to the theme of ontogeny, or what is more commonly called *development*, while *evolution* refers strictly to the higher-level process that Haeckel called phylogeny.

Stem cells are purported to have the unique ability that upon dividing, one of the "daughter" cells and its lineage can differentiate into specialized cell and tissue types (by expressing specific genes), while the other remains in the non-specialized stem state and is able to produce both lines of specialized and further stem cells. Once a cell line has begun to specialize, it is not normally able to revert back to the prior undifferentiated (or more accurately, less differentiated) stem state, though the property of being a "stem cell" is one of degrees. Totipotent mammalian stem cells are capable of creating all the different cell types in a developing embryo, including those that will form the chorion and placenta that are not strictly speaking part of the new individual animal, but part of the support system required to help it grow in the mother's womb. The group of cells known as the "inner cell mass" that occupy the interior of the early ball of cells (or blastocyst) in human embryogenesis are called *embryonal stem cells*. These, and stem cells found in many organs and tissues of adult mammals and other animals, such as the liver or bone marrow, are pluripotent, meaning they are capable of producing both renewed cells of that specific tissue type and further pluripotent stem cells. It is because these pluripotent stem cells have the ability to create or regenerate new healthy tissue and organ cells that many are hopeful that scientists will be able to use them to treat people suffering from organ damage or other diseases. Recent research has shown that even some adult somatic cells (those that have differentiated to become specific tissue types) can be "induced" by experimental intervention to become stem-like. These induced pluripotent stem cells (or iPSCs) are frequently spoken of as being "reprogrammed," and will be discussed further in Chapter 8.

While the topic of stem cells is fascinating and their application in research and in biomedical therapy raises many important social and ethical questions, the metaphor itself is perhaps quite straightforward. It expresses the basic causal notion of one thing being derived from another, as when we say, for instance, "It all stems from the fact that . . ." Yet it is notable that it evokes a distinctly organic image of causality that is drawn from the source domain of nature as opposed to human agency or machines.

Cells, Race, and Gender

A classic illustration of how societal values can influence how scientists interpret the objects of their studies involves the study of fertilization – that is, the union of the gamete cells in sexually reproducing organisms. In a now classic paper "The egg and the sperm: how science has constructed a romance based on stereotypical male–female roles," the anthropologist

Emily Martin looked at the language used to describe the respective roles of the egg and sperm cells in the events of fertilization. Since Aristotle, women have been portrayed (and expected to behave) as passive, while men are expected to be the active agents of political and economic affairs. Martin argues that these societal stereotypes have been projected onto the gamete cells of even non-human animals. (It is worth noting just how natural it seems to us to think of the egg and sperm cells themselves as gendered – the "female egg" and the "male sperm" – when we would hardly think of a liver or brain cell in such terms. In fact, as the developmental biologists Scott Gilbert and Clara Pinto-Correia point out, the terminology of "gamete" cells derives from the Greek for marriage partners, and hence we speak of fertilization as a "union" between the two cells.) The egg, so the typical story went, emerges from the ovary to be passively swept along (like so much debris) through the fallopian tube, while the sperm aggressively and vigorously swim their way toward the object of their mission, the "penetration" of the waiting egg. In another classic paper on this topic, Scott Gilbert and co-authors remark on the similarity between these traditional accounts of fertilization ("sperm sagas" they call them) and the heroic tales of Homer's *Odyssey* or Virgil's *Aeneid*.

The egg is sometimes described as if it were Sleeping Beauty waiting on Prince Charming to awake her from her slumber by the magical effect of his kiss. Martin discussed research in the 1980s that presented evidence of the egg's active role in the process. The egg, it turns out, is not so much "penetrated" by the sperm; rather, the sperm is entrapped by the egg and pulled in by means of tiny microvilli on the zona pellucida, a glycoprotein coating that surrounds the egg. (As a matter of fact, the egg's passivity in fertilization had been challenged much earlier, in the 1930s, by the developmental biologists Ernest Just and Frank Lillie.) Martin notes that while sperm and egg bind to one another in a typical protein receptor and ligand fashion (recall the lock-and-key analogy), the researcher who made the discovery reversed the standard practice of description and deemed the sperm-borne receptor (which is typically construed as the passive partner in the lock-and-key metaphor) the very active-sounding "egg-binding protein," while the ligand (ZP3) found on the egg surface is called a "sperm receptor." Martin offers this as an example of how deep-seated prejudices can bias scientists' ability to see the world objectively even when they are attempting to correct past errors.

A similar story is recounted by the neuroscientist Meg Upchurch and women's studies professor Simona Fojtová with respect to the portrayal of glial cells. Long understood to be responsible for creating the myelin sheath that insulates neurons or for providing them with a supportive scaffolding, glial cells were typically described as subservient in importance to the neurons that create and transmit the electrical–chemical signals of the brain and central nervous system ("glia" comes from Greek for *glue*). But when it was later discovered that some glial cells known as astrocytes also communicate with one another and with neurons through calcium ions, and help to regulate the formation of synapses (the important communication junctions between neurons), characterization of their activity revealed interesting biases about the worth of certain kinds of traditionally gendered labour. Initially glial cells were said to perform "housekeeping" and "nursemaid" duties in support of the executive functions of the neurons, which were frequently portrayed as the real "stars of the show." As recognition of their significance in the neural hierarchy rose, so too did their job descriptions, eventually being promoted to the status of "masters of the synapse" and "architects of memory," reflecting the more highly esteemed occupations associated with traditionally male-dominated professions.

While not strictly concerned with cells, historian Nancy Leys Stepan has argued that in the nineteenth century metaphor helped to perpetuate sexist and racist attitudes in Western society by analogizing non-whites to the feminine. (It needs to be noted that Stepan's argument really applies to analogy proper as she does not provide any instance of an actual metaphor to illustrate the point.) The analogy linking race and gender, she writes, "served as a program of research." It could do this, she claims, because the metaphor launched a whole series of what Max Black called "a system of associated commonplaces," in this case prejudices against women and non-whites held by white males of the scientific class. The pseudo-scientific theories of race, from the time before and after Darwin's theory of evolution, flourished because the people who promoted them could so easily cherry-pick data to fit the thesis that women and non-white males were inferior. That is, they could take what they regarded as positive analogies between the two separate classes and use them as motivation for finding which of the apparently neutral analogies could be forced into the positive bin as well.

Stepan writes that, "The confusion of metaphor for reality in science would be less important if metaphors did not have moral and social consequences in addition to intellectual ones . . . Metaphors shape our perceptions and in turn our actions, which tend to be in accordance with the metaphor." This concern falls under what Kostas Kampourakis has called "the use of bad metaphors and the bad use of metaphors," the remedies for which he recommends are that scientists be explicit about the limits of the metaphors they employ, that they are in fact metaphors with potential heuristic benefits at most, and that they be conscious of how non-scientists may and do interpret them.

I agree with these remarks, with the one amendment that we need to recognize that language that is originally metaphorical is not precluded from eventually being accepted as literally apt and even as expressing literal truth, as the history of the terms *cell* and *cell factory* attest. But this may provide even more reason to employ metaphors with caution and forethought, or as is so often said, with "eternal vigilance." The duty to use scientific metaphors responsibly falls not only on scientists, but on all of us: science communicators, journalists, academics, students, and citizens included. As the philosopher Andrea Sullivan-Clarke and others have argued, a promising way to ensure scientists do not fall prey to the influence of "ingrained analogies" is to promote the social inclusion of and respectful critical engagement with under-represented groups in science and the academy at large.

6 Evolution

Natural Selection, the Tree of Life, and Selfish Genes

While other scientific-technological developments may have had greater material impacts on how we live (for instance, the unleashing of fossil fuels to drive the industrial revolution, atomic energy, or the creation of digital computers), none has had a greater impact on how we understand what it means to be human and our place in the universe than the Darwinian theory of evolution. Darwin was not the first to propose that humans and other species have not always existed in their present forms, and that they have gradually developed or emerged from earlier forms of life. Theories about the "transmutation" or "transformation" of species were proposed in Europe by members of his grandfather's generation, including Lamarck, Geoffroy Saint-Hilaire, and Darwin's own paternal grandfather Erasmus. In his own time, people like Robert Chambers and Herbert Spencer wrote popular and philosophical essays espousing what was frequently called the hypothesis or principle of *development*. (The concept of *evolution*, so familiar to us today, originally referred to a specific thesis about embryogenesis and only later acquired its current meaning.) Darwin's chief contributions, as published in 1859 in *On the Origin of Species*, were: (1) to synthesize a vast amount of evidence in favour of the thesis that species are not immutable; (2) that new species gradually develop from what can be recognized as subspecies varieties under the right environmental conditions; and (3) to propose a hypothesis for the mechanism by which new species emerge. While people commonly speak of "Darwin's theory," it is important to recognize it is actually a composite of at least three separate theses: (1) the mutability of species; (2) the community of descent of all extant species from one or possibly a few original forms; and (3) natural selection as the prevalent mechanism of change.

In the development and articulation of his ideas, Darwin relied substantially on analogical reasoning. And this analogical reasoning was in turn reliant on several key metaphors. One was "the Tree of Life," to represent the thesis of the community of descent or shared ancestry of all species. The other concerned his hypothesized mechanism for species transmutation, which he called "natural selection," a choice of terminology based on an analogy with the process of artificial selection practised by humans in the production of new varieties of domesticated plants and animals. Natural selection as a mechanism for species change has been intimately tied to the competitive "struggle for existence" and the "survival of the fittest." As the historian and philosopher of biology Michael Ruse has written, "starting with the key notions of selection and struggle themselves, Darwin embedded his discussion in the practices, events, and metaphors of his day, his British industrial culture" which was characterized by fierce economic competition, exploitation of the working class by capitalist entrepreneurs and wealthy investors, division of labour, and technological innovation. Doing so not only allowed Darwin to communicate persuasively with his cultural peers, the metaphors also allowed him to think about the problems of species' origins and adaptation in tractable ways, and to create solutions that were themselves strongly fashioned by metaphorical language and imagery. The metaphors were therefore of significant heuristic value, but also, as we will see, they help to provide the cognitive objectives of explanation and understanding. But are they separable from the cultural metaphors and context of Darwin's time – or of our own? I will return to that particular question at the end of the chapter.

Darwin's ideas – or what soon came to be known simply as Darwinism (despite the contribution of the naturalist Alfred Russel Wallace) – created a splash in the placid surface of Victorian society and the waves it created have continued to slosh back and forth ever since. But despite detailed discussion of empirical evidence and experimental facts upon which the Darwinian theory of evolution is founded and confirmed, the real reason it has proven so controversial is that it involved an exchange of fundamental metaphors for understanding what it means to be human and of our place in nature. Prior to Darwin's publication of *On the Origin of Species* and his later books, the basis for Western civilization was the traditional Judeo-Christian metaphor that humans were the manufactured products or artifacts of a supernatural intelligent creator. This is vividly displayed in Paley's famous analogical argument for the existence of God wherein the human eye and other organic structures are likened to a mechanical watch. Darwinism threatened to replace that image with an alternative metaphor that humans are the natural and relatively recent offshoot (a twig among many, on a branch among many) of an ancient and expansive tree of life forms that has grown up from the soil or seas of the planet over vast ages. For many people, accepting this metaphor means accepting that we are not the crown jewel in some intentionally designed piece of craftsmanship, and that we are literally kin with all the other "lowly" species with which we currently share the planet. It's no wonder that the most stalwart critics of evolution are not swayed by the presentation of scientific evidence or fact, when for them it is really a battle of fundamental metaphors.

But in order to understand both the scientific and the broader philosophical and religious aspects of evolutionary biology, we must understand these key metaphors rightly, what their positive contributions are, and why they may stick in some people's throats or otherwise impede further progress.

Evolution Is a Metaphor

Perhaps it is best to begin by noting that the very term that has come to stand for the whole field of science in question is itself a metaphor. *Evolution* comes from the Latin verb *evolvere,* which means to unroll, as in unrolling a rug or scroll. The term featured in a debate that played out in the seventeenth century concerning embryogenesis. One school of thought, known as preformationism, held that the young animal (human or otherwise) must already be contained in miniature within the fertilized egg, so that the process of development was akin to unrolling a rug to reveal in larger form the new being. This idea was vividly represented in a drawing by Nicolaas Hartsoeker in 1694 of a homunculus (or little man) within the head of a sperm cell (those who, like Hartsoeker, believed the preformed individual was contained in the sperm cell were called "spermists," while "ovists" opted for the egg). While this might seem absurd, consider the difficulty of defending the alternative account, known as epigenesis, which maintained the fertilized egg begins as an apparently homogeneous minuscule mass, nothing like the fully developed child at birth, but somehow (quite mysteriously) all the bones, muscles, and organs appear gradually therefrom. Science today recognizes elements of truth in both accounts. From what *appears* to be a homogeneous fertilized egg cell, a baby with differentiated tissues and organs does gradually develop, but this is possible because each of the daughter cells resulting from repeated acts of division contain the genetic "information" required to build the specialized structures, and the genes are initially activated by transcription factors that the mother has asymmetrically deposited into the egg. In other words, the new human is in a sense already "preformed" in the genetic "blueprint" of the zygotic nucleus, but requires the external epigenetic inputs of maternal and environmental factors to initiate and complete development.

It is because the concept of evolution was closely associated with the subject of embryogenesis that Darwin and others like him who were interested in the question of the origin of species did not use the term in relation to that subject. As mentioned above, Darwin and other naturalists tended to speak of the transmutation or transformation of species, or development, which is confusing because today that term is restricted to processes of embryogenesis and regeneration of lost and damaged parts, the subjects of developmental biology. In the German-speaking regions of Europe, naturalists like Ernst Haeckel tended to use the term *Entwicklung* (or the older form *Entwickelung*), which covered both the creation of new organisms and new species. Or in Haeckelian terms: ontogeny and phylogeny, which he saw as intimately and causally linked.

But the notion of evolution carried with it the preformationist thesis that the end result of the process was preordained or necessitated by the beginning stage (of the egg and/or sperm). When applied to the question of the origin of species, therefore, "evolution" would suggest an inevitability in the historical process, with humans perhaps being the ultimate goal or *telos* toward which the whole process moved. Some thinkers were happy to replace the older idea of a *scala naturae* or Great Chain of Being with humans occupying the top rung with this more historical and dynamic account in which God did not make humans and other species all at once in their current forms but allowed them to slowly evolve or develop. But not Darwin. Once he had become convinced that species are not immutable, Darwin's observations on the geological and geographical distribution of species around the globe convinced him that the process by which new species emerge is not predetermined, but highly contingent upon a number of factors. These included local environmental conditions, changes over geological time to the environment, the migration of plants and animals into new environments, and the availability of variation in the phenotypic structure and behaviour of organisms.

It is for this reason that originally Darwin referred to his theory not as the theory of evolution, but "descent with modification." However, although it is frequently noted that Darwin did not use the term "evolution" in the first edition of *The Origin*, it is worth noting that it closes with this now rather famous sentence (with my emphasis added): "There is grandeur in this view of life, with its several powers, having been originally breathed into a few forms or into one; and that, whilst this planet has gone cycling on according to the fixed law of gravity, from so simple a beginning endless forms most beautiful and most wonderful have been, and are being, *evolved*." Evidently the meaning of the term was itself undergoing subtle change.

Natural Selection

To explain how it is possible for a new species to emerge from one previously existing, Darwin proposed the idea of natural selection. He was not alone in thinking of the basic idea, as Alfred Russel Wallace struck upon it independently, though later than Darwin, and others had noted it in more restricted applications before him. In some ways it was an ingeniously simple idea; however, it required a complex of separate ideas to line up properly (like the pins of a combination lock) for it and its significance to become apparent.

Darwin's close observation of individual organisms of the same species (he spent nearly 10 years studying all living and extinct species of barnacle prior to publishing *The Origin*) had taught him that within any population there is frequently a great deal of variation of traits and behaviours if one looks closely enough. No two plants or animals of the same species are exact carbon copies of one another, not even parent and offspring, although they are likely to be pretty close. A chance reading of the political economist Thomas Malthus' *Essay on the Principle of Population* impressed upon him that the fecundity of a species tends to exceed the resources required to see every plant seed or young animal live to maturity and reproduction. Far more dandelion seeds are produced on a lawn than could find space, water, nutrients, and sunshine to thrive were they each to successfully reproduce. Even a slowly reproducing animal like the elephant would soon exhaust available resources if every young calf born in a herd managed to reach sexual maturity and have young of its own. Darwin realized that this would create a competition among the members of the same population of plant or animal species for these scarce resources: space, food, shelter, mates, etc. And because there is variation of traits and habits among these competing individuals, some of these variants are likely to provide a competitive advantage to those fortunate enough to possess them. These individuals can be expected, on average, to be more successful in surviving to sexual (or asexual) maturity and reproducing. And if these advantageous traits tend to be passed on to their offspring, the population as a whole will over time come to be dominated by individuals in possession of these better-adapted traits.

The effect, in other words, will appear *as if* nature, or the local environment to be more specific, had *selected* those organisms with specifically well-adapted characteristics to be the most successful at thriving and reproducing its kind. That in a nutshell is the idea of natural selection. It will look *as if* the organisms thriving in any environment were intelligently and purposely designed to succeed in that habitat and that particular way of life. But it would require no intelligence, no forethought, no ultimate plan, and no supernatural creator. It would be the result of an entirely natural process. Darwin used the practice of how animal and plant breeders improve their stock by selectively breeding individuals with preferred traits as an analogy to help the reader see how nature itself could over time produce organisms well adapted to their specific environments. "I have called this principle," he wrote, "by which each slight variation, if useful, is preserved, by the term Natural Selection, in order to mark its relation to man's power of selection."

Darwin devotes the first two chapters of *The Origin* to providing evidence of the natural variation that exists in populations of plants and animals in the wild and under domestication. This variation is the raw material from which nature will "select" the most adapted types. But in order for nature to select preferentially the better-adapted or fitter types, it cannot be the case, he believed, that every plant or animal, no matter the variation in trait it possesses, is able to thrive and reproduce. In order for evolution to occur (i.e., for there to be a change over time in the dominant properties and behaviour of the population), he believed there must be a competition so that on average the better-adapted types have greater reproductive success. (Evolutionary biologists today note that all that is required for the population to undergo a shift in allele frequencies or phenotypes is that there be a genetically based variation in reproduction rate among organisms – they don't necessarily need to compete for resources at all.) But Darwin's belief in the necessity of competition led him to emphasize the "Struggle for Existence" in nature. This has been misconstrued by many to mean that Darwin was suggesting that nature rewards only the most aggressive and brutal behaviour, an inference that has been used to justify violence, military aggression, and the "every man for himself" attitude of "social Darwinism" and other ideologies opposed to social welfare programs. Yet Darwin himself was very clear how he intended to use the notion of struggle:

> I should premise that I use the term Struggle for Existence in a large and metaphorical sense, including dependence of one being on another, and including (which is more important) not only the life of the individual, but success in leaving progeny. Two canine animals in a time of dearth, may be truly said to struggle with each other which shall get food and live. But a plant on the edge of a desert is said to struggle for life against the drought, though more properly it should be said to be dependent on the moisture. A plant which annually produces a thousand seeds, of which on an average only one comes to maturity, may be more truly said to struggle with plants of the same and other kinds which already clothe the ground ... As the mistletoe is disseminated by birds, its existence depends on birds; and it may metaphorically be said to struggle with other fruit-bearing plants, in order to tempt birds to devour and thus disseminate its seeds rather than those of other plants. In these several senses, which pass into each other, I use for convenience sake the general term of struggle for existence.

There need not be a literal struggle between two organisms, though there could be, in order for differential reproductive success to occur. Nor does nature necessarily favour non-cooperative or anti-social behaviour. If an animal that tends to behave cooperatively toward its peers successfully leaves behind more progeny than a more selfish variety of the same species, then in

Darwin's terms, the cooperative variety will have won the competition and will be selected as the more favourable variation. (Note that this does not mean the less well-adapted variety – the "loser" in the competition – will fail entirely to reproduce or be completely "weeded out." For evolution by natural selection to occur, all that is required is that a variation in trait or behaviour becomes more frequent in a population because of the reproductive advantage it provides over other existent variations in the same population.)

Darwin's suggestion that nature *selects* or prefers certain traits over others was a metaphor intended to highlight the analogy to the practice of agriculturalists, horticulturists, and those who breed pigeons, dogs, or horses as a hobby. The positive analogy between the two systems is that in both there is a distinct tendency for the frequency of available variation to move in a particular direction: toward those traits that appeal to the fancy of the human breeder in the one system, and toward those traits that confer a reproductive advantage (by whatever means) in the other.

Darwin quite shrewdly noted an important negative analogy too, but turned it toward his advantage. For surely someone will object that when humans select among those plants or animals to breed, they pay attention to only a very few specific features: a softer fur, more colourful plumage, or a more docile demeanour, whereas nature has no such specific preferences in mind. Indeed! Darwin remarks,

> It may metaphorically be said that natural selection is daily and hourly scrutinizing, throughout the world, the slightest variations; rejecting those that are bad, preserving and adding up all that are good; silently and insensibly working, *whenever and wherever opportunity offers*, at the improvement of each organic being in relation to its organic and inorganic conditions of life.

It is true that nature operates in a way so much broader and more penetrating than humans as to count as a significant disanalogy – and yet for that very reason, Darwin remarks, this demonstrates why natural selection ought to be recognized as a power "as immeasurably superior to man's feeble efforts, as the works of Nature are to Art." Darwin seems to be saying to those who wish to see the wonders of nature as reflections of a superior and ingenious force, you can have your cake and it eat too!

But this points to the existence of another negative analogy, and one that proved more problematic precisely because many people are inclined to take it as a positive analogy. This is the implication that natural selection involves a teleological principle, that nature operates analogously to a conscious and goal-seeking human agent to bring about some preferred end. In an attempt to defuse this misconstrual, Darwin adopted in the fifth and final sixth editions of *The Origin* a suggestion from the philosopher Herbert Spencer.

Survival of the Fittest

Actually, it was Alfred Russel Wallace (who independently hit upon the idea of natural selection) who recommended Darwin use Spencer's phrase "the survival of the fittest" to avoid the confusion over purposiveness suggested by the term natural selection. To say that nature selects or favours certain traits or organisms over others is a useful means of setting up the analogy Darwin wished to draw between natural selection and artificial (or human) selection. But the metaphor implies that, like conscious human beings, nature is an agent with its own distinct goals and plans in accordance with which it directs events. Darwin wrote in the fifth and sixth editions, "I have called this principle, by which each slight variation, if useful, is preserved, by the term Natural Selection, in order to mark its relation to man's power of selection. But the expression often used by Mr. Herbert Spencer of the Survival of the Fittest is more accurate, and is sometimes equally convenient." Here, unfortunately, Darwin was wrong. For although "survival of the fittest" does avoid the suggestion of teleological agency on the part of nature, it was misleading in a whole other way, and has perhaps created an even greater obstacle to people's understanding of the theory of evolution. The phrase is misleading in at least three key ways.

For one, it introduces the idea of "fitness," which is an additional metaphor with the potential to mislead. Darwin did previously use the notion of an organism or a trait being well or poorly "fit" to a particular habitat, and in doing so he was invoking a spatial analogy (later to be developed as the metaphor of an ecological niche, which will be discussed in the next chapter). Darwin spoke of an organism or trait "fitting" its habitat or mode of life in analogy to how a key fits a lock (although he does not explicitly voice that comparison). For instance, he writes that "each being assuredly is well fitted for its place in nature"; that "an intermediate variety will often be formed, fitted for an intermediate zone"; plants are "fitted for extremely different climates"; a species is "fitted to some peculiar line of life"; or "fitted to its new home." Darwin employed another spatial metaphor in early letters, notebooks, and in the first edition of *The Origin* that confirms this interpretation of how he understood the language of "fitness." He frequently employed the analogy that "The face of Nature may be compared to a yielding surface, with ten thousand sharp wedges packed close together and driven inwards by incessant blows, sometimes one wedge being struck, and then another with greater force." In accordance with this analogy, successful organisms and species would be those that "fit" well and deeply into this crowded surface, so that they were not easily displaced by other newcomers. But it also invites the question, "Who is swinging the hammer that is driving these wedges into the metaphorical surface of Nature?" Here again, the metaphor misleadingly suggests some teleological agency lies behind Darwin's conception of natural selection.

Second, because "fitness" is a metaphor it is open to multiple interpretations. One of the most common is the inadequate and misleading interpretation that, as it often means in relation to humans, to be fit means being athletic and muscular. And because many sports are very physically competitive and even aggressive (e.g., wrestling and boxing), people frequently think that to be fit in an evolutionary sense means to be aggressive and even violent. But recall that in Darwinian terms fitness is only about successful reproduction, and that is just as readily achieved (among animals) by being social, cooperative, friendly, and altruistic. And of course plants can only be described as aggressive in a metaphorical sense.

Third, the phrase "survival of the fittest" suggests that *only* the fittest individuals survive, and subsequently reproduce, which is not at all what Darwin or later evolutionary biologists believe and know to be true. The essential point that Darwin sought to capture with the term "natural selection" is that of differential reproduction, that individuals with variations that are more adaptive relative to the local environment are likely to leave behind more progeny of that type than others with less adaptive variation. "Survival of the fittest" suggests that only the fittest or most adaptive survive, let alone reproduce. So, while natural selection allows for the continued existence of less well-adapted variation (albeit in fewer numbers), survival of the fittest suggests a drastic reduction in variation.

This has also encouraged the further confusion that Darwin's proposed mechanism for evolutionary change (natural selection under the supposedly equivalent expression "survival of the fittest") is an exclusively negative force that destroys or "weeds out" maladaptive traits. But Darwin knew, and biologists today have shown even more, that natural selection can also be a positive or constructive force. Variation in traits is said to arise "randomly" or "by chance," by which is meant that variation occurs independently of whether it will prove to be beneficial, deleterious, or neutral in its effect. Phenotypic variation arises from various events at the genetic level (e.g., gene mutation, recombination and crossing-over of chromosomes, gene shuffling resulting from sexual reproduction, epigenetic tagging of chromatin), all but the last of which are said to be "non-directed" because this variation happens in "all directions" not only or even preferentially toward fitter, more beneficial variation. But natural selection is non-random in the sense that organisms in possession of beneficial traits (those that happen to confer upon them an advantage within that specific environment) are likely to leave behind more offspring (i.e., enjoy greater reproductive success). Once natural selection begins to increase the frequency of adaptive traits in a population, it will also have the effect that more individuals with this and other adaptive traits will breed together, further concentrating and amplifying the adaptive variations in the population. Hence, natural selection is not simply or exclusively a destructive or eliminative force, but also a "creative" one; but the phrase "survival of the fittest" does not capture that feature at all. And it continues to create confusion for people's understanding of evolutionary biology to this day. For these reasons it is probably better that biologists, educators, and science communicators avoid and discourage the use of the phrase "survival of the fittest" altogether.

Darwin responded in later editions of *The Origin* to the objections and misinterpretations made by some readers regarding his use of the metaphor of natural selection. With just a hint of frustration detectable in his tone, Darwin wrote:

> In the literal sense of the word, no doubt, natural selection is a false term; but whoever objected to chemists speaking of the elective affinities of the

> various elements? – and yet an acid cannot strictly be said to elect the base with which it in preference combines. It has been said that I speak of natural selection as an active power or Deity; but who objects to an author speaking of the attraction of gravity as ruling the movements of the planets? Everyone knows what is meant and is implied by such metaphorical expressions; and they are almost necessary for brevity. So again it is difficult to avoid personifying the word Nature; but I mean by Nature, only the aggregate action and product of many natural laws, and by laws the sequence of events as ascertained by us. With a little familiarity, such superficial objections will be forgotten.

(One can almost hear him mutter under his breath, "Come on, people, seriously!") And yet there is still another metaphor central to the Darwinian evolutionary account of biology.

The Tree of Life

In Darwin's time, many people assumed that the relationships among the various living forms were best represented by the classical Aristotelian–Christian idea of the *Scala Naturae*, the "Ladder of Nature," or Great Chain of Being. This is a linear and hierarchical classification of beings, ranging from minerals at the bottom, through plants, to simple animals, ascending to humans, with angels and God at the pinnacle or top step. Each form of life or species was, in theory at least if not in completed practice, assigned a step or rank in this hierarchy according to a very anthropocentric measure of perfection or complexity. Early evolutionists like Lamarck modified this image of a static ladder by suggesting a moving escalator, whereby simpler animals like worms can transform over time into rudimentary vertebrates like snakes or fish, which in turn develop legs and develop further into monkeys, apes, and eventually humans. Lamarck, however, posited separate chains of transmutation for the plant and animal kingdoms, and assumed that, among the animals at least, there was a final goal toward which these processes were aiming. An inner striving for perfection drove worms and other "lower" animals to improve themselves. It is no coincidence that these early ideas of upward mobility gained traction among the *philosophes* in the years leading up to the French revolution of 1789. To account for the continued presence of the lowly organized worms, protozoa and other infusoria (including bacteria), Lamarck posited the spontaneous generation of new lines of life, which would in their turn seek to climb the ladder of species transmutation.

Whereas Lamarck invoked an image of one species of plant or animal transforming into another (creating a linear series like links in a chain), the diagram that appeared in his *Philosophie Zoologique* of 1809 depicting the relationship between the various classes of animals includes a series of dots that hints at an upside-down branching tree. Several other writers in the eighteenth and early nineteenth centuries also broke from the chain metaphor, opting instead for networks or tree-shaped systems to classify the similarities of form and function in living or fossil plants and animals, but these were not evolutionary systems.

Once he became convinced of the mutability of species, Darwin opted for a process of divergence, whereby one species does not wholly turn into another to create a chain, but splits into two or more, like the branching of a tree. Noting that there are often identifiable varieties within a given species of plant or animal (e.g., the different breeds of domesticated dog or varieties of apple trees), Darwin argued that under the influence of natural selection these subspecies groups could over time diverge from one another far enough to count as wholly new species. The idea that the process by which new species are originated might take on a tree-like form occurred to Darwin in 1837, and was recorded in his Notebook B on the transmutation of species (Figure 6.1). Twenty-two years later, the sole image included in *The Origin* depicts this branching pattern of divergence to illustrate his conception of how incipient varieties may gradually branch out to become good species of their own, perhaps to produce further divergent branches or to come to an end in an event of extinction (Figure 6.2).

About this branching pattern of speciation and divergence, Darwin wrote:

> The affinities of all the beings of the same class have sometimes been represented by a great tree. I believe this simile largely speaks the truth. The green and budding twigs may represent existing species; and those produced during each former year may represent the long succession of extinct species. At each period of growth all the growing twigs have tried

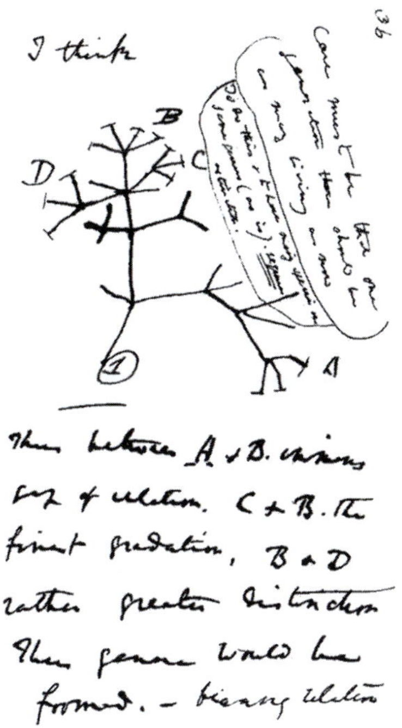

Figure 6.1 Charles Darwin's "Tree of Life" notebook sketch (public domain).

to branch out on all sides, in the same manner as species and groups of species have tried to overmaster other species in the great battle for life. The limbs divided into great branches, and these into lesser and lesser branches, were themselves once, when the tree was small, budding twigs; and this connexion of the former and present buds by ramifying branches may well represent the classification of all extinct and living species in groups subordinate to groups.

The Tree of Life was a particularly apt metaphor to capture the current relationship among extant species (using the analogy of a genealogical or family tree of relations), but it also vividly captured the dynamic and historical

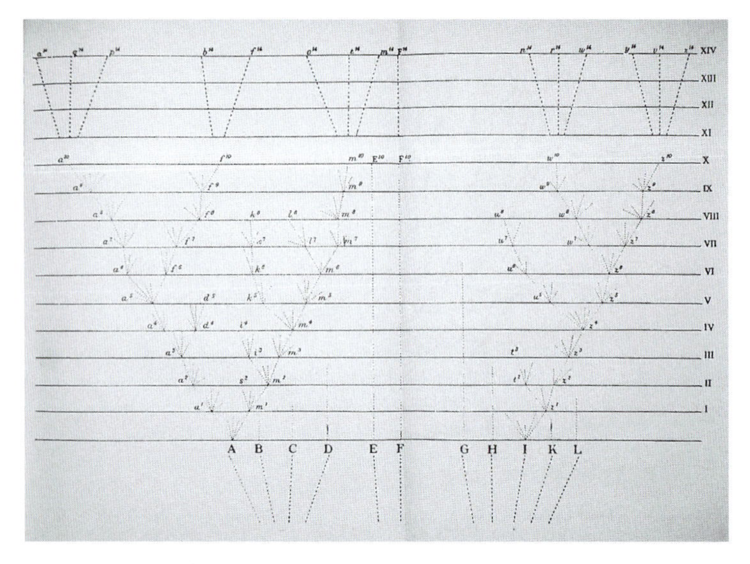

Figure 6.2 Tree of Life (Darwin, C. (1859). *On the Origin of Species by Means of Natural Selection or The Preservation of Favoured Races in the Struggle for Life*. First edition. London: John Murray (public domain)).

development of those relationships via the image of the growth of a tree with many branches, some of them dying and producing no further growth, while others remain vibrant and continue to bud off new shoots which will eventually become branches in their own right. Reflecting on the implications of the tree metaphor, Darwin said, "The terms used by naturalists of affinity, relationship, community of type, paternity, morphology, adaptive characters, rudimentary and aborted organs, &c., will cease to be metaphorical, and will have a plain signification."

Darwin's tree diagram inspired the German zoologist Haeckel to draw his own series of phylogenetic trees depicting the evolutionary histories of various groups of animals and of life in general. But where Darwin's image employed a minimalistic line sketch with a tree-like structure, Haeckel, who was a gifted artist, used tall and sturdy oak-like trees as his model (Figure 6.3).

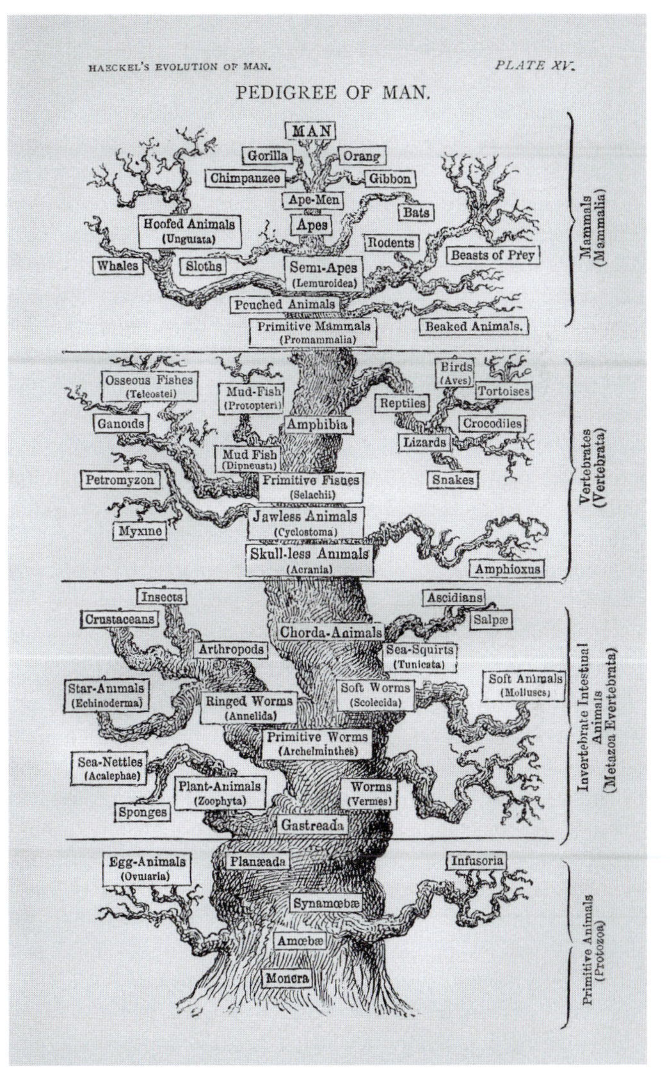

Figure 6.3 The stem tree of man (Haeckel, E. (1897). *The Evolution of Man: A Popular Exposition of the Principal Points of Human Ontogeny and Phylogeny*. New York: Appleton & Co., plate xv. (public domain)).

Note that despite the accurate realism of Haeckel's trees, several conventions used in their construction are discernible. First, they all lack leaves, unsurprising since the point is to represent the historical relationships of common descent, and this is handled sufficiently by the tips of each branch or twig counting as a species or higher taxon. Haeckel's trees are also unlike Darwin's original diagram in that they invoke a hierarchical scale, according to the morphological or physiological characteristic of interest. Whereas each currently existing species in Darwin's diagram reaches the top of the page, Haeckel's trees reflect his commitment to an "idealist morphology" according to which some organismal forms are more "highly evolved" than others. So, for instance, single-celled and amorphous amoebae lie at the bottom of the tree as the most primitive form of life, and as we ascend the tree the forms become more complex. (Figure 6.3 has often been submitted as evidence that Haeckel held the anthropocentric and teleological assumption that "man" is the ultimate goal toward which evolution strives, but this is easily debunked by inspection of any of the other trees he drew for different animal classes. When he drew similar trees for other types of animals, for instance, the coelenterata or zoophytes which include corals, anemones, and jellyfish, they are featured at the top and centre of the tree. Figure 6.3 is, as one can easily see, titled the "Stammbaum des Menschen" or "Stem-tree of Humans," not the "Tree of Life.")

The tree metaphor has two other significant implications: (1) That once branched, lines of descent never merge, in analogy with how the branches and stems of a bush or tree (almost) never fuse together; and (2) that the tree representing the evolutionary history of all life on earth – the Universal Tree of Life, as it's frequently called – has one common trunk or source, which would presumably represent the original or most primitive form of life, the universal common ancestor from which all later forms and species of life have descended. But are these two analogies between the two systems positive or negative analogies? For a long time, they were assumed to be positive, when in fact they ought to have been regarded as neutral analogies, because it was simply not known whether distinct species can combine to form one or whether life had one origin or a universal common ancestor. Scientists using the tools of molecular biology have lately come to question both of these assumptions.

Molecular Biology Threatens to Uproot the Tree of Life

The "molecular revolution" in biology brought with it knowledge of the structure of DNA and its role in the process of inheritance, the origins of novel phenotypic variation via gene mutation, chromosome rearrangement, and sexual recombination, and the amino acid composition of proteins; it also opened a window into a whole new world of evolution at the molecular level. Scientists could now explore evolutionary change not only in populations of organisms, but also in the sequence of nucleotides in a species' genome and in the sequences of amino acids making up its proteins. Francis Crick was one of the first to realize that these were potentially rich sources of information about a species' evolutionary history and its taxonomic relationship with others. Assuming that nucleotides in the DNA randomly mutate at a relatively consistent rate, and that gene sequences of special importance to an organism's survival would tend to be highly conserved, scientists could use candidate sequences as "molecular clocks" to estimate how long it had been since two species last shared a common ancestor. In other words, the historical record mined from molecular data could be used to draw up phylogenetic trees. This suggestion was taken up by the chemists Linus Pauling and Emile Zuckerkandl, who investigated the difference in amino acid sequences in the hemoglobin protein of humans and chimpanzees. Among the things they found was that these two were more similar to one another than either was to the hemoglobin of orangutans, suggesting that humans and chimpanzees are one another's closest "evolutionary cousins."

Carl Woese, a microbiologist, followed this lead and in the early 1970s began studying the ribosomes – the "machines" that make proteins – of diverse types of bacteria. The ribosome was a good choice of molecular clock because it is an essential component of all cells, both prokaryote and eukaryote, occurs by the tens of thousands in each cell, and consists of several units of RNA in addition to protein. Woese focused on the bacterial small subunit known as 16S ribosomal RNA (16S rRNA), which is similar to the 18S rRNA found in plants, animals, and fungi. Through painstaking and laborious work, Woese eventually concluded that the accepted idea that there are two chief kingdoms of life based on cell structure – the eukaryotes that have a nucleus and the prokaryotes that do not – was meaningless. Woese had found a whole group of bacteria whose ribosomal profile was so unique and unusual that they were as different from the other prokaryotes as either was to the eukaryotes. He insisted they deserved their own status as a separate third kingdom, which he called the *archaebacteria* (or ultimately *Archaea*, based on the assumption that they were more ancient because they tended to live in environments of such extreme heat, acidity, and salinity as seemed fitting of the first cells to emerge on an inhospitable early planet earth).

But further study of the molecular profiles of Bacteria, Archaea, and Eukarya revealed several other big surprises. For one, it provided firm evidence that the chloroplast in plants (an organelle responsible for photosynthesis) and the mitochondria in animals (an organelle characterized as the "power plant" responsible for ATP synthesis) are both descendants of once-independent bacteria that somehow at some distant time came to live inside a larger unicellular organism. This is known as the endosymbiotic theory. Both mitochondria and chloroplasts still possess a small number of their own genes, including those for the small subunit rRNA, inspection of which confirms that chloroplasts are the descendants of cyanobacteria and mitochondria are related to the alpha-proteobacteria. Further sequence evidence suggests that, in fact, the Archaea and the Eukarya share a more recent evolutionary ancestor, meaning that from the "root" of the universal Tree of Life, the Bacteria and Archaea split early, with the Eukarya branching off from the Archaea at a later time. More surprising still is the discovery of bacterial genes in the nuclear genomes of eukaryotes. Given the endosymbiotic origin of chloroplasts and mitochondria from bacterial ancestors, it is not surprising that some bacterial genes may have found their way into the nuclear genome, but one would expect any such to be involved in respiration or photosynthesis. This, however, is not the case. The microbiologist Ford Doolittle and his lab have shown that genes of crucial function in the eukaryote genome have not been inherited "vertically" in a typical pattern of evolutionary descent from an earlier ancestor, but "laterally" or "horizontally" in a fashion suggestive of more recent accumulation by some process like endosymbiosis or perhaps by means of a viral vector (viruses sometimes take up genetic material from the host cells they infect and then introduce this foreign DNA into subsequent victims; in fact, this is a method commonly used in genetic engineering). Similar evidence of not very tree-like horizontal transfer of genetic material has been found in Archaea and Bacteria.

While it has been known for some time that bacteria "swap" genes in this non-vertical fashion (it is one of the reasons that antibiotic-resistant bacteria have become such a problem), its wider occurrence throughout and between the three kingdoms has led Doolittle and others to call into question the "consensus view" of the universal Tree of Life. Doolittle wrote in 2000, "we must now admit that any tree is at best a description of the evolutionary history of only part of an organism's genome. The consensus tree is an overly simplified depiction." So what would be a more accurate picture? Doolittle suggested that while most of the portrayal of evolutionary relations among the eukaryotes (animals, plants, and fungi) would continue to look mostly tree-like, with fusion lower down between eukaryotes and bacteria where mitochondria and chloroplasts were obtained, below that there would be a tangle or reticulated pattern of branch fusions among the bacterial and archaeal domains; and most strikingly, any attempt to identify a precise and unique root or main trunk would be arbitrary. Woese had also argued that rather than a single universal common ancestor, as Haeckel and others had supposed, the base of the tree would become a tangled network representing what would have been a community of early microbial cells that regularly exchanged genes, like a community of music or software enthusiasts freely sharing files (Figure 6.4).

Some scientists have argued that while the bifurcating tree metaphor is a useful heuristic for thinking about and analyzing data for some groups, it is really only a special case of a more fundamental network metaphor or pattern. The possibility of new species forming from the hybridization of existing species, now known to be common among plants, is a further reason to recognize that "branches" can in fact merge. As the network advocate David Morrison notes, a genealogy will be tree-like in the case that there are no reticulations, otherwise all "trees" are really networks. And yet, as Doolittle has remarked, it is important to note that in these discussions one must recognize that sometimes what is being portrayed are species lineages (species B and C sharing a recent common ancestry through species A), while at other times

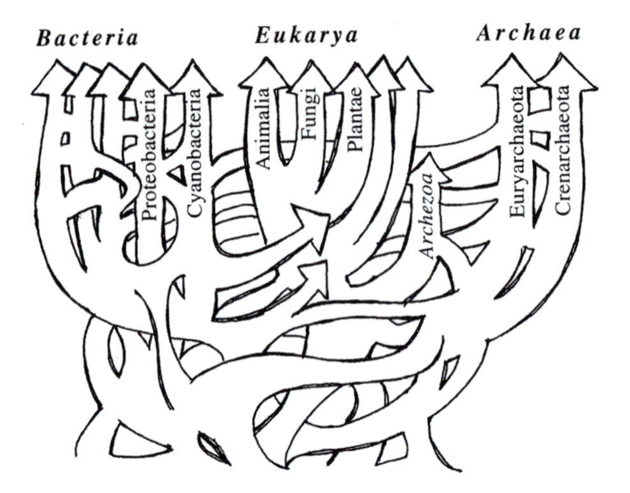

Figure 6.4 The Tree of Life with lateral gene transfer (Doolittle, W.F. (1999). Lateral gene transfer and the universal tree of life. *Science* 284(5423), fig. 3. Reprinted with permission from AAAS).

the concern is with the history of genomes (as in events of endosymbiosis when two genomes merge), or it may concern just a few genes or even one nucleotide sequence (as in the case of horizontal gene transfer). Attempts to portray such diverse sorts of events in a single diagram can be misleading for those unfamiliar with the scientific details. It's complicated because depending on what molecular evidence one is looking at, one can construct very different "trees," "networks," or some combination of the two.

Just how potentially misleading all this talk about revising the Tree of Life can be was illustrated in 2009 when the magazine *New Scientist* featured on its front cover the declaration that "Darwin was wrong: Cutting down the tree of life." What exactly was Darwin supposed to have been wrong about? The mutability of species? The efficacy of natural selection to create new species? That all forms of life share a common history of descent with modification? Or simply that the tree-like diagram he printed over 150 years ago would prove to be fully accurate forever more? Many complained that the cover invited people to draw any or all of these conclusions. The accompanying articles beyond the cover made clear it was only the last claim that was under dispute. But is it really a surprise that he was unable to anticipate every new fact and feature about molecular evolution unveiled by scientific techniques and concepts he could not have been expected to know anything about? The attention-grabbing headline was used (quite predictably) by religiously inspired critics of evolution to insist that the whole edifice – or more aptly the whole tree – of materialistic Darwinism was rotten to the core. But this again just speaks to how variously metaphors can be interpreted and to the potential risk that they may lead to considerable misunderstanding, especially once set loose into the popular conversation.

It should have come as no great surprise to anyone that the "Tree of Life" is not a literal tree but a metaphor, and that like every metaphor will entail both positive and negative analogies. Ford Doolittle said as much in 1999 when he wrote that "The tree of life is not something that exists in nature, it's a way that humans classify nature." The key issue is whether the metaphor leads to important insights, questions, and discoveries. On that score there is no question that the tree metaphor has borne much fruit. But we must also ask whether a metaphor captures accurately and adequately the facts of the target domain to which it is applied. Here, the answer seems to be that it is generally suitable for one of the major domains of life, the Eukarya where the branching pattern of diversification is real, but less so for the other two, the Archaea and the Eubacteria. In light of the fact, however, that most of life on earth is, and has ever been, of those two microbial domains, it is understandable that some strongly insist the significance of the "Tree of Life" imagery ought to be diminished in future discussion and representation of life's evolutionary history.

The Selfish Gene

One of the best-known metaphors concerning evolutionary science in recent times is undoubtedly the "selfish gene" – the anthropomorphic conception promoted by Richard Dawkins in his 1976 book of that title. In 2017 the Royal Society voted Dawkins' *The Selfish Gene* the most influential science book of all time, surpassing even Darwin's *On the Origin of Species* and Newton's *Principia Mathematica*. In addition to being highly influential, the selfish gene metaphor has also been one of the most misunderstood. It illustrates how a metaphor chosen with greater regard for its rhetorical attractiveness than for its ability to provide cognitive insight can be misleading and counterproductive.

Dawkins' purpose in writing *The Selfish Gene* was to show how evolutionary theory could account for altruistic behaviour in animals, including humans, when natural selection might be expected to eliminate individuals who engage in acts that lower their own reproductive fitness or chances of reproducing. Certainly we consider it morally laudable when a human or other animal risks its own life to save another, but from the perspective of natural selection it makes little sense to do so. And if we suppose that, at least among non-human animals, behaviour is strongly influenced (if not wholly determined) by instinct rather than conscious deliberation, then we would expect that the allele or alleles implicated in such behaviour would quickly die out with those who carry them. Altruism therefore appeared to pose a significant challenge for the Darwinian theory of evolution by natural selection.

Dawkins set out to show how by taking a gene-level view of this and other phenomena, the Darwinian account could make perfect sense of things. The theoretical account he presented was not his own, but had been developed by William D. Hamilton and was known as kin selection or inclusive fitness theory. The idea was that if an individual gave its life so that a close relative survived, although the genes in the altruist would be lost, because a closely related kin is likely also to carry the allele for altruism it would survive in them and have an opportunity to reproduce. Dawkins' contribution was to help emphasize in particularly vivid language that what appeared to be *selfless* behaviour at the level of the organism was, in fact, *selfish* behaviour at the level of the genes. "Selfish," at least in the sense that what appeared to be of benefit for something other than the hypothesized allele in question was really to its benefit.

Dawkins sought to highlight how taking the gene's-eye perspective could provide explanations of organismal behaviour that were consistent with the role of natural selection as the chief mechanism for the evolution of life on earth. Even otherwise competent biologists had fallen into the mistake of saying that plants or animals engaged in various behaviours and reproductive strategies in order "to ensure the survival of the species" or for the benefit of the group to which they belonged. This cannot be right, Dawkins insisted, following the lead of George C. Williams in this case, because neither species, groups, nor even individual organisms for that matter are reproductive units. Species do not reproduce themselves, they do not have offspring, nor do groups of animals. And even sexually reproducing plants and animals are evolutionary dead-ends, because they do not replicate themselves entirely, but only pass on one-half of their genomes to the next generation (that goes for us humans too). The only thing that truly gets replicated is DNA, of which genes are made, and so it is only genes that can be the true beneficiaries of the process of natural selection, Dawkins argued. The gene's-eye perspective was intended to turn normal thinking on its head: Just as it has been said that a chicken is only an egg's way of making more eggs, complicated organisms like us are only DNA's way of making more DNA. According to Dawkins, we humans, just like all other plant and animal bodies, are actually only vehicles that have been created by means of natural selection because they have proven beneficial to the reproduction of DNA.

The "selfish gene" metaphor was Dawkins' convenient shorthand label for this gene-centric view of evolution. But the apparent anthropomorphic ascription of conscious goal-seeking agency to DNA created a morass of misunderstanding and confusion for many people. It was not surprising that those who had not read the book concluded that Dawkins was saying that humans are determined to behave in selfish ways because we are controlled by genes that are themselves selfish, or that some people are selfish because they possess the "gene for selfishness." This was not Dawkins' intent at all. But even those who read the book could be forgiven for concluding that Dawkins was suggesting something like this when they got to the end of Chapter 2, where he wrote that the once free-living genes of ancient pre-cellular times now "swarm in huge colonies, safe inside gigantic lumbering robots, sealed off from the outside world, communicating with it by tortuous indirect routes, manipulating it by remote control. They are in you and in me; they created us, body and mind; and their preservation is the ultimate rationale for our existence . . . and we are their survival machines."

Dawkins later said that he could just as easily, and with perfect consistency to the gene-centric perspective, have called his book "The Cooperative Gene." The genes of a single genome of course must "cooperate" with one another to help build a healthy, functioning organism if they are to have any reproductive success. Dawkins used the analogy of a team of individual rowers who must coordinate their strokes perfectly with one another if they are to win the race. What Dawkins meant in describing genes as selfish is that the various alleles of a particular gene are in competition with one another in any given population. For instance, suppose there are alleles that are involved in the production of fragrant chemicals that happen to attract insect pollinators. Different allelic versions of these scents will then be in competition with one another for the insects' attention, but naturally they must cooperate with all the other genes in their own plant genome if they are to have any chance of being rewarded by natural selection. Dawkins chose the "selfish" adjective to emphasize that in every case in which natural selection is involved, all that matters from the gene's perspective – it being the proper unit of reproduction – is whether it gets replicated or not. An individual gene may benefit from the fact that every gene in the genome gets reproduced – just as a rower benefits if every member of her team wins the race – but that is incidental to what really matters, namely its own success. Analogously, no *competitive* athlete who plays a team sport really wants their team to win without them, otherwise what would be the point of their being on the team?

Selfish DNA

In addition to the idea of selfish genes, there is also the concept of selfish DNA. This, by contrast, is analogous to the horribly competitive team member who only cares about their own success, the athlete who refuses to pass the puck because they just want to win the individual scoring title, even if it means their team fails to make the playoffs. So-called selfish DNA refers to bits of genetic material that have hit upon, simply by chance and coincidence, behaviour that leads the cell's replicative mechanisms to make more copies of it than it normally would, or even should. Selfish DNA is akin to a virus or cancerous cell within the genome and may ultimately end up being destructive for both the organism and the DNA itself. But for a while it will enjoy some success at differential reproduction within the genome, if not in the external population of organisms. Because this sort of selfish behaviour can be destructive, natural selection has favoured cells that have happened upon means of "quality control" over their DNA replication. Alternative names for selfish DNA include: ultra-selfish genes, parasitic DNA, and genomic outlaws.

Junk DNA

Some selfish DNA manages to get itself replicated in the cell, even though it does not appear to code for any protein, ribosomal or transfer RNA, or have any identifiable regulatory function. Non-coding DNA is popularly called "junk DNA," and by some estimates comprises up to 98.5 percent of the human genome. It appears as segments of meaningless repeated nucleotide sequences that have been randomly "cut and pasted" throughout an organism's genome. Some of this "junk" could be selfish DNA or vestigial remnants of ancient genes now rendered defunct by mutation, but having no seriously deleterious effect. There is debate, however, about whether so-called junk DNA might have a non-protein-coding regulatory function. The ENCODE project, standing for Encyclopedia of DNA Elements, has claimed that more than 80 percent of the human genome is functional on the basis that some biochemical activity associated with the non-coding regions has been discovered. Critics like Ford Doolittle have insisted these claims are based on a confusion between biochemical activity and biologically adaptive function. Doolittle and others insist that ascription of function in biology should be restricted to structures and processes that have been selected for, not just those that may have incidental effects. Philosophers of science like to point to this as a good example of how science can benefit from the kind of conceptual analysis that philosophy specializes in. It may also reveal again how much can ride on the choice of a metaphor. As the old saying goes, one person's junk is another's treasure.

Spandrels, Functions, and Adaptation

A disagreement about biological function was also at the heart of another important metaphor introduced by the evolutionary biologists Stephen J. Gould and Richard Lewontin in their 1979 paper "The Spandrels of San Marco and the panglossian paradigm: a critique of the adaptationist programme." Gould and Lewontin were critical of those they called pan-adaptationists who, they claimed, assumed *prima facie* that every feature of an organism served some functional purpose and therefore required an explanation invoking natural selection to explain its existence. Gould and Lewontin appealed to the architectural feature known as a spandrel to make their point that a feature of an organism may exist simply because it is a non-functional by-product of some other adaptive feature that has been selected for. When in the design of a classic building, like St. Marks Basilica (San Marco) in Venice, two arches are placed side by side, a triangular space is created at the top near the ceiling. These spaces, known as spandrels, are often decorated with beautiful mosaics. Were one not careful, one could draw the mistaken conclusion that they were specifically designed for the purpose of holding the artwork, when in fact they are simply the result of a constraint imposed by the arches which are designed for a function. The point was to caution evolutionary biologists not to assume that any trait in which they happen to be interested (such as the human chin, or the biochemical activity of non-coding DNA) is an adaptation rather than a possible spandrel.

The paper sparked a lively and long-lived debate about the criteria by which one identifies adaptive functions and the distinction between the function a trait may currently have and the reason it evolved, which may be for a quite different function. Gould called a trait that originally evolved for one purpose but now serves a separate and distinct one an *exaptation*, whereas a spandrel is strictly speaking a non-functional by-product of some distinct selected adaptation. The discussion generated by the spandrels paper also highlighted the dangers of conceptually breaking whole organisms down into atomistic parts whose separate functions can then be discerned, in analogy with how we study the various parts and mechanisms of a clock, computer, or other designed artifact. While *similar* to machines in some regards, living organisms, we must remember, are their own category of thing.

We have seen now just how central metaphor has been to the conceptualization and development of the science of evolution, and I have left out quite a number of other important instances, such as adaptive and epigenetic landscapes with fitness peaks and valleys, and selective pressures and forces, to name only a few. It would seem difficult after all this to imagine evolutionary science without the metaphors, or that it might even have been possible without them. What, then, does this mean for its scientific status?

Is Evolution All Just a Metaphor Then?

Michael Ruse argues that if not for the cultural tradition of viewing the plant and animal kingdoms as creations designed for some purpose, Darwin would not have asked the types of questions he did (nor Wallace), and would not have eventually arrived at the answers he did. The metaphors, Ruse insists, are constitutive of and integral to the science. But what, then, does that suggest about its objectivity? He writes:

> One has to transcend dichotomies of objective/subjective, discovered/ created, description of reality/social construction. Science, Darwinism in particular, falls on both sides of the divides. Today's evolutionary biology is clearly subjective, in that through its metaphors it reflects the culture in which it was formulated – Western (especially British and American), industrialized, mass agricultural, Christian, militaristic, sexual, and much more. We could not have had the theory had we not been living in a Judeo-Christian type of society, asking about origins and about humans and so forth. The Greeks did not ask these kinds of questions, and they were not evolutionists. The same goes for notions of struggle and selection and so forth. On Andromeda, say, there might be a society of intelligent scientists who simply don't find our kinds of questions that interesting. Perhaps some metaphors are universal – you cannot have intelligence without the ability to make artifacts, and this spells design. But, generally, the members of such an alien society would not be evolutionists; they would certainly not be Darwinians, because they would frame their questions in different ways.

Ruse speaks here to the importance of framing, language, and metaphor to the human activity that is science. Even when it strives to be objective – to provide a "view from nowhere" as the philosopher Thomas Nagel described it – and even when it comes closest to succeeding, it ultimately must still provide a view or account that makes sense to us humans, not to Andromedeans or other celestial intelligences.

So, yes, the metaphors of evolutionary (Darwinian) theory may frame the phenomena to be explained and arrange the explanations for their existence in ways that reflect the historically contingent experiences of human beings on the third closest planet orbiting a rather average star, just one among billions. But to regard that as a major flaw in the science is a bit like complaining of a road map that the route you are currently driving on is not actually green and the landscape is far from flat. The point of the map is to help you navigate from point A to B, and if it does that, so long as we recognize the conventions used in its creation, that is all that really matters. Likewise, one might argue, if we are careful to recognize the element of subjective perspective introduced by metaphors and that we should not confuse them for the Objective God's Eye View of Reality, what matters is whether the science helps us to navigate and to make sense of (in human terms) the particular slice of reality to which it is meant to be applied. Understandably, not everyone will be satisfied with this instrumentalist or pragmatist response. The realism–antirealism debate in philosophy of science has typically focused on the problem of whether or not we can claim to have knowledge of unobservable hypothetical entities and processes (like atoms or gravitational waves), but science's reliance on metaphorical language presents a whole other and as yet under-appreciated issue.

But we turn now to the science of ecology and some of the metaphors that have helped define and shape it.

7 Ecology

The Balance of Nature, Niches, Ecosystem Health, and Gaia

Darwin's evolutionary view of life emphasized change over stasis. The older creationist account, meanwhile, supposed that species of plant and animal had been specially designed to fit perfectly the environments in which they live, and there was little reason therefore for them to change at all. Aside perhaps from the intermittent earthquake, volcanic eruption, massive flood, or other rare catastrophe, they were assumed to live in a stable environment. But Darwin had been convinced by Charles Lyell's uniformitarian account of geology that the earth is in a constant state of change, and as a consequence species will have to adapt if they are to maintain a good fit to their surroundings. Over long periods of time, new islands rise up from the depths of the ocean due to volcanic activity, or sink as the ocean floor shifts; mountains rise and fall; periodic cooling and warming of the planet causes ice sheets and glaciers to expand and contract, leaving behind new lakes or deserts. So species evolve, migrate, or go extinct. And one cannot forget that each plant, animal, or other living thing is dependent on many other species for its daily survival, so that as each population of organisms is molded by natural selection to better fit its surroundings, it too is a part of the environment of all the other species whose success or failure is entangled with its own, creating a dynamic and ever-changing network of interdependent relations. Because organisms are parts of one another's environments, when they change, so too must all the others who are part of that complex web.

In modern ecological science these relations of biotic (living) and abiotic (nonliving) interdependence are collected under the familiar term "ecosystem."

Questions about the degree to which ecosystems are stable or undergo constant change (and at what scale) have been a perennial source of debate. Darwin himself, although he championed the idea of evolutionary change in nature at large, also seemed to endorse the idea that an established system of relations among species of plants and animals reflects a "balance of nature." Today, public debate over terminology such as "global warming" or "climate change" illustrates how these duelling visions of an environment in stasis or in constant change continue to influence our understanding of nature and life on the planet.

Modern ecology is a rich science with a diversity of metaphors to match, so some selectivity in this chapter is necessary. I have chosen to discuss several metaphors that have been or currently are influential and that touch on the issue of stability versus change. The term "ecology" itself has metaphorical origins, drawing on the earlier metaphorical expression of the "economy of nature," which referred to the contribution made by plant and animal species to the supposed stability of nature's balance. Darwin's metaphor of a species or variation "fitting" its environment was later developed by ecologists into the concept of an ecological "niche," which in its original form suggested a position or role in the economy of nature pre-existing a species and just waiting to be filled by it or some other candidate organism. This was eventually replaced with the idea that organisms do not simply find a niche waiting for them, but rather "construct" the niche by transforming the environment through their presence and behaviour. Organism and environment are then conceived as being in a dynamic process of mutual co-construction.

Questions remain, however, about the nature of changes created in an ecosystem due to the migration or introduction of new "invasive" species and the potential extirpation of "native" species. Inevitably judgments about the nature of such changes will draw on anthropocentric values falling into the categories "good" and "bad," and have resulted in debate about whether a metaphor like "ecosystem health" is a legitimate tool for monitoring and achieving desired applied science outcomes, or whether it is too subjective and value-laden to be a respectable scientific concept. Such discussions raise the perennial question of humans' place in nature – are we a part of it or above it? Does scientific objectivity require an account of nature free of human values, or is such an account an unrealistic, impossible, or possibly irrelevant goal? Lastly, if we take a global perspective on the geological, chemical, and biological processes that are occurring daily on the planet, is there a legitimate scientific basis for describing the earth itself as a living organism or system, as the Gaia hypothesis does? The difficulty of retaining "objective neutrality" on these issues – especially given their political, economic, and societal significance – illustrates again how pressing is the question of whether science is supposed to include or exclude traces of the human beings who conduct it and the values that motivate our efforts. This is at the heart of the topic of metaphor and science. We begin with the metaphor of the economy of nature.

The Economy of Nature

Carl Linnaeus, the founder of modern taxonomy and author of the *Systema Naturae* (1735), also published in 1749 a treatise on "The Oeconomy of Nature" which emphasized the contribution made by each species to the balanced relations of exchange in the natural world. "Economy" derives from the ancient Greek *oikos*, meaning "household," and *nomos*, meaning "law." Sixteenth-century works written on the "animal economy" and "plant economy" referred to the organization and orderly function of living bodies. "Economy" was used metaphorically to refer to the harmonious arrangement or organization of parts supporting vital functions. Linnaeus, among others such as Sir Kenelm Digby, extended this principle of balanced relations to nature as a whole. The functional and reciprocal balance between the fungi and maggots that decompose the bodies of dead plants and animals, so that they may become resources contributing to the continued cycle of life, were considered by Linnaeus to have been established by God. Each species occupied a particular station in the economy of nature according to its function in maintaining the harmonious persistence of God's creation.

Darwin also invoked the economy of nature in *On the Origin of Species* and in his other writings, largely in keeping with this earlier usage. But because of his reliance on the Malthusian principle of competition, many commentators have agreed with Friedrich Engels' judgment that "The whole Darwinian theory of the struggle for existence is simply the transference from society to animate nature of Hobbes' theory of the war of every man against every man and the bourgeois economic theory of competition, along with the Malthusian theory of population."

Setting aside the question of how damning Darwin's reliance on Malthusian political economy is for the validity of his science, it remains the case that he used the economy of nature metaphor and cognates (e.g., economy of growth) essentially to refer to the relationship of interdependence among various species of plants and animals. More relevant is how the metaphor of the economy of nature was adapted and developed by later thinkers into what is now the science of ecology.

Ecology

Ecology, the term and the science, was introduced by that prodigious coiner of neologisms Ernst Haeckel, Darwin's ardent German colleague. Ecology (or *Oecologie* in Haeckel's original German), stems from the Greek "oikos" (house) and "logos" (an account of or study of, in modern usage). In the second volume of his massive work *Generelle Morphologie*, published in 1866, Haeckel introduced ecology as "the collective science of the relations of organisms to their environment." Metaphorically, this would be the study of how the household of nature is managed – that is, how individual organisms manage to "make a living" in the environmental surroundings in which they find themselves. Haeckel no doubt was following the path laid down by Darwin, who spoke of an organism having a "line of life" in the economy of nature similar or analogous to a "line of work."

Others continued to develop the science of ecology through an anthropocentric lens, notably Karl Möbius who in 1877 introduced the idea of an ecological "community" and Józef Paczoski who coined the term "phytosociology" to describe a particular approach to the study of plant ecology in 1896. Economic analogies were used by Ralph Lindeman, who in 1942 characterized the various contributions made by different species and forms of life into the two chief categories of "producers" and "consumers" (broadly outlining plants and animals; a third category largely consisting of fungi and bacteria is known as decomposers).

Frederic E. Clements, an influential plant ecologist of the first half of the twentieth century, regarded an ecological community of different plants as a kind of "supra-organism" that progresses through regular developmental stages, analogous to the development of an individual plant or animal. Clements proposed that after a disruption of an existing plant community (due to fire, perhaps, or human disturbance such as agricultural practices), if left to its own natural development, a period of "colonization" would occur during which particular types of annual herbs would establish a patchy "pioneer community," followed gradually by a succession of ground-cover perennials, shrubs, and small trees, until finally reaching a "climax" of larger woodland forest capable of reproducing itself.

Ecological Niche

While ecology has clearly been shaped by metaphors drawn from human social experience as a source, one of its best-known concepts is borrowed, like the notion of a spandrel, discussed in the previous chapter, from the domain of architecture. The physical and functional "space" that an organism occupies in an environment or ecosystem is called a "niche." A niche is defined by the *Oxford English Dictionary* as a shallow recess in a wall in which an ornament may be displayed (Figure 7.1). The term was first used in its ecological sense in 1910 by R. H. Johnson and further developed by Joseph Grinnell in an article of 1917 titled "The niche relationships of the California Thrasher." The term is originally derived from the middle French word *nicher*, meaning to nest, so the architectural use of the term is itself originally a metaphor sourced from biology or the natural world. It was eventually defined by the ecologist Charles Elton in terms of the additional metaphor of a *food chain*, which he used to characterize the hierarchy of relations between organisms in their potential roles as consumer and consumed – for example, grass is consumed by rabbits, which are consumed by foxes, which may in turn be consumed by eagles.

Recall that Darwin had spoken of an organism occupying a "place" in its environment corresponding to its "line of life" in the economy of nature, analogous to a "line of work" by which it survived and reproduced. Competition from other organisms with similar lines of life or occupations, he believed, drives species to adapt to other spaces in nature's economy where competition is less intense, and this would lead over time to the divergence of new species from the parent species (driving the new branches

Figure 7.1 A literal niche. Niche with a sculpture by Antoine Coysevox, in the Les Invalides, Paris (Creative Commons Attribution Share-Alike 3.0 Unported licence; https://commons.wikimedia.org/wiki/File:Paris_-_D%C3%B4me_des_Invalides_-_Statue_-_PA00088714_-_003.jpg).

in the Tree of Life further apart to use another of his key metaphors). Darwin's comparison of nature to a surface into which thousands of wedges are being driven is also captured by the notion that each species occupies – or seeks to occupy – its own niche in which it can thrive.

The concept of an ecological niche has been very influential, but has also been developed in at least two fundamentally distinct ways. One version construes the niche in largely spatial terms as the physical or environmental space in which an organism lives; this is often known as a "habitat niche." Another treats the niche in terms of the organism's role in or contribution to the ecological community, and is known as a "functional niche." Niches can be operationally defined in terms of abstract geometrical spaces wherein various biotic and abiotic factors relevant to the species are measured and represented for further mathematical analysis.

A common assumption in earlier thinking about niches was that, like a physical space or an occupation in an economy, niches can exist separately from the organism that occupies it. This is an analogy quite evidently suggested by the metaphor's source because a literal niche is an empty space that may or may not be occupied by a statue. Another assumption that appears to have been naturally suggested by the metaphor was that similar niches could exist in different environments or regions, and that a niche occupied by one species in one environment could be filled by a different species in another. For example, in the case of Darwin's finches on the Galapagos Islands, one species of finch is described as occupying a niche typically filled by woodpeckers on the mainland, and presumably before this species began to exploit this niche by feeding on insects that live on trees and under the outer bark, the niche remained vacant.

These may have originally been taken as positive analogies between the two systems linked by the metaphor, but it was eventually realized that they ought to be regarded as neutral analogies requiring proper investigation. Many scientists have subsequently concluded they are in fact negative analogies. As Richard Lewontin explains, "the use of the metaphor of a niche implies a kind of ecological space with holes in it that are filled by organisms, organisms whose properties give them the right 'shape' to fit into the holes." In fact, Lewontin describes how the language of a species adapting to its environment belies the assumption that an organism's environment exists independently and prior to the organism and its behaviour. It is the organism that must be molded or adapted to fit the environment. However, Lewontin and others have urged that the metaphors of adaptation and niche need to be replaced with one of *construction* in recognition of the organism's active modification of its abiotic and biotic surroundings.

So, according to niche construction theory, organisms are actively constructing and manipulating the relevant aspects of the physical space in which they live into conditions more suitable for their survival and reproductive success. They do not just passively step into an empty niche they find waiting for them, but actively shape and construct an environment, "carving out" a niche for themselves. Obvious examples include the drastic engineering projects executed by beavers, damming streams to turn fields into ponds or lakes. But even the smallest and most inconspicuous plant is modifying its physical and chemical environment through its presence and continued growth both above and beneath the soil. Darwin wrote an entire book (his last) about how earthworms help create the very soil in which they and countless plants and other small creatures live by digesting small particles of rock and excreting the modified remains, furthermore loosening and aerating the soil in the process of creating it. In Lewontin's pithy expression, "Just as there can be no organism without an environment, there can be no environment without an organism." Although it seems natural to suggest that, by some understanding of the idea of an environment we'd like to say there is an environment on Mars even if the planet is lifeless, that organisms and niches are in a continual process of co-construction or co-adaptation seems more plausible.

While Lewontin provided the impetus for this shift in metaphorical perspective, it has been more fully developed by the biologist John Odling-Smee, who coined the term "niche construction." The upshot is to emphasize that the environment and the organism are not independent factors of which the former responds to the latter; rather, they are coupled, meaning that through their modification of their environments, organisms are actively influencing their own evolution. Another implication drawn by advocates of niche construction theory is that organisms do not only inherit DNA from previous generations of their species, but also an environment that in some cases will prove advantageous to their survival and reproduction, whereas in others may be more challenging. For instance, some trees, such as white pines, in adulthood create conditions of low light that are inhospitable to their own seedlings but suitable for the growth of hardwoods. There is some debate about the extent to which niche construction therefore qualifies as a factor separate from, but of equal evolutionary significance to, natural selection. Regardless of the ultimate conclusion of this debate, niche construction theory highlights that organisms and their environments are dynamic systems in a constant state of change, even if it is difficult for us humans to notice it at the time scale limits set by our own average lifetimes. Expanding the range of our senses and understanding of the world is one of the key dividends paid by scientific inquiry.

In their efforts to understand how organisms and their environments change and influence one another over time, ecologists have both broadened and deepened the scope of their analyses to include a more systemic account of the geochemical factors and thermodynamic cycling of energy and nutrients that drive biological communities. This involves the adoption of the now familiar *ecosystem* approach.

Ecosystem

The concept of an ecosystem was introduced in Arthur Tansley's 1935 article "The use and abuse of vegetational terms and concepts." Combining the *oikos* metaphor with the conceptualization of organisms and their environment as a holistic system of integrated components, the ecosystem concept has deeply influenced the science of ecology. Clements had earlier referred to a community of plants and animals living within a particular environment as a "biome." Tansley broadened the scope of analysis to include the abiotic factors so as "to include with the biome all the physical and chemical factors of the biome's environment or habitat – those factors which we have considered under the headings of climate and soil – as parts of one physical *system*, which we may call an *ecosystem*, because it is based on the οίκος or home of a particular biome." It is with this eco-systemic approach that ecology became popularly associated with a holistic (or wholistic) philosophy that is rhetorically contrasted with the reductionism supposedly characteristic of the physical and natural sciences (e.g., particle physics, chemistry, molecular biology), in accordance with which objects, processes, or phenomena are to be understood by breaking them down metaphorically and/or literally into their more basic parts and their respective properties.

Beginning in the 1960s and 1970s the terms "ecology" and "ecological" became synonymous with the popular environmental movement that tended to be motivated by a desire to preserve "nature's balance," which was perceived to be under threat from human social and economic impact. That ecosystems naturally tend to exist in states of equilibrium was a common belief among ecologists. Tansley, for instance, in introducing the ecosystem concept, wrote that "In an ecosystem the organisms and the inorganic factors alike are *components* which are in relatively stable dynamic equilibrium. Succession and development are instances of the universal processes tending towards the creation of such equilibrated systems." Tansley was himself critical of the idea promoted by Clements and others that a biotic community was significantly analogous to a living organism, with its own developmental stages and a tendency to maintain homeostasis (a term introduced by the physiologist W. B. Cannon in 1926 to denote the ability of a living system to maintain its physical state relatively constant despite changes in its external environment). Yet Tansley agreed that once established, an ecosystem will tend to maintain and preserve itself in dynamic equilibrium if the impacts and insults to which it is subject do not exceed its limits of tolerance.

In fact, the belief that nature exhibits a tendency to maintain a harmonious balance between various opposing forces, like the population sizes of predators and prey, is longstanding and widespread, and is perhaps summed up best through the ancient Chinese idea of Yin and Yang, the two opposing forces that create and maintain the world. The balance-of-nature metaphor has been a seductive and powerful influence on both popular thought and ecological science, but as the ecologist Kim Cuddington notes, it is not value-neutral and implies that "nature is orderly and beneficent, perhaps in the same way that the, now absent, divine presence had similarly been beneficent."

The Balance of Nature

In the West, the balance-of-nature metaphor has theological origins arising from the assumption that a creator would ensure balance and harmony in its creation, much in the way that a human artist or artisan seeks aesthetic and functional harmony or balance in the objects they design. Pre-modern medicine was similarly based upon the premise that good health is conditioned upon the balance of the four bodily humours.

But the conservation biologist Brendon Larson draws upon Lakoff and Johnson's conceptual metaphor theory to suggest that our embodied experience as bipedal organisms provides a potentially deeper source for the balance metaphor's attractiveness. We naturally associate good health with our ability to stand erect and maintain our balance on two feet, while many pathologies and dangerous situations can throw us off balance. Regardless of how intuitive the experience of balance may be as an analogy for understanding other systems, the thesis that ecological systems tend to maintain a natural balance has come under criticism as of late. John Kricher, in his extended analysis of the idea, *The Balance of Nature: Ecology's Enduring Myth*, suggests the belief is an illusion created by a "scale effect." If one focuses at the right scale, for instance, at the relation between populations of hares and foxes, or at the passage of recently disturbed plant communities through stages of weedy pioneers to a more stable and self-reproducing forest of trees, one can draw the conclusion that natural ecosystems seek to maintain certain stable relationships or states. But if one expands one's perspective, he insists, one will find that the overwhelming trend is of dynamic continual change. Slow and gradual movement of the planet's tectonic plates results in continental shift and radical redrawing of global maps, the rise and fall of mountain chains, and land-bridges between islands and mainland. Migration of plant and animal species into new territory creates new environments and radically alters existing ones. Around 2.4 billion years ago, anaerobic bacteria altered the originally methane-rich atmosphere in what is known as the great oxygen catastrophe and created an opportunity for the selection of aerobic bacteria that could thrive in the presence of this poisonous gas, making possible eukaryotic multicellular life forms like us after the chance merger of several bacterial and archaebacterial cells. In our hubris, we humans lay claim to being the greatest threat to nature, when in fact we are probably only the latest greatest threat after the single-celled bacteria that unwittingly engaged in geoengineering of a truly global scale.

The notion of the balance of nature, Kricher writes, "has always been a fuzzy, poorly defined idea that nonetheless has had great heuristic appeal throughout the ages because it seems so self-evident"; yet it is of "little value within evolution and ecology," because it lacks clear definition (which is arguably true of all metaphors). While many ecologists, *qua* professional scientists, now question the adequacy of the balance-of-nature idea, there are obvious reasons why many people, *qua* human beings whose range of concern is narrowed to comparatively shorter time frames measured in decades, or perhaps a century or two at the most, worry about the environmental changes they witness around them. And so ecology, as the most obvious environmental science, has come to be equated with environmentalism and critiques of the dominant political, social, and economic systems that are implicated in the many ecological crises of our age.

Current discussions about global warming reveal the tension between those who advocate for adaptation to what is now already inevitable climate change brought on by our past behaviour and some for whom it seems nothing less than doing whatever is required economically and societally to stop it altogether is acceptable. Some it seems would prefer that we prevent any further change to natural ecosystems or even return to pre-industrial environmental conditions. But if critics of the balance-of-nature thesis are correct, the choice between adaptation to changing environments and freezing time are false options. We have no choice but to adapt to change regardless of its origins; we cannot preserve nature as though it were a museum exhibit. But this is not to say that we cannot make significant changes to our politics, economics, and daily lifestyles that will make the sort of environmental conditions to which we must adapt less threatening and inhospitable.

"To some non-scientists, however," writes Michael Allaby,

> "ecology" suggests a kind of stability, a so-called "balance of nature" that may have existed in the past but that we have perturbed. This essentially metaphysical concept is often manifested as an advocacy for ways of life that are held to be more harmonious or, in the sense in which the word is now being used, "ecological." The idea is clearly romantic and supported by a somewhat selective view of history, but it has proved powerfully attractive.

On the other hand, those who are opposed to making changes to the existing economic and political system, for instance, by reducing our reliance on fossil fuels, are too quick to make facile criticisms that environmentalists would have us all move back into caves. Sound public policies will need to be premised upon sound visions of reasonable and acceptable relations between humans and the rest of nature.

It is unhelpful to argue over which changes and how much is "natural" or damaging to the "natural balance" of things. This would require a fair and objective assessment of what such a "natural" state of affairs in the absence of humans would be. But to insist on the separation of humans from nature is to continue down the same path that led to our current environmental problems. And yet to concede that humans are a part of nature is not to immediately grant that any change we bring about is *ipso facto* "natural" and therefore "good" or beyond scrutiny. That would be to continue making the naturalistic fallacy that just because something is natural it is morally blameless or even praiseworthy. Infant mortality due to easily preventable infectious illness is entirely "natural" in the sense that it will and does occur if we choose not to prevent it, but few I hope would therefore suggest that vaccines and access to clean water are bad ideas.

The matter of scale – largely temporal – is key. Skeptics of the anthropogenic roots of climate change are surely correct that "The climate is always changing," but this is as irrelevant as stating the undeniable fact that "People are always dying" in response to easily preventable health crises like gun violence or the COVID-19 pandemic. No one reasonably wants to die before what is the natural time scale for members of our species. And similarly, there is no reason why we as a global society should not attempt to delay or outright prevent the predictable collapse of the climate and environmental systems on which *we* rely, even if it is an undeniable fact that the planet and nature as a whole will continue to change regardless of our efforts and will ultimately become inhospitable to our continued existence when the sun becomes a red giant in roughly five billion years.

So we must find ways of talking about and understanding our relationship to the rest of nature that are both factually accurate and practically effective. And that probably means adopting the right metaphors. Which brings us back to the point that science is not only or always just about attaining a factually or literally true description (to hold a "mirror to nature"), because while that may accurately (or not) describe what is known as "pure" science, we also expect science to solve for us more "practical" problems, what is known as "applied" science. And for the purpose of applied science, metaphors can serve as very effective "tools" even if they might not serve as entirely objective "lenses." If we want to avoid sawing off the branch of the tree on which we sit, then we may need to adopt figures of speech that we can collectively understand and work toward to achieve the result we desire. In his book *Metaphors for Environmental Sustainability: Redefining Our Relationship with Nature*, the conservation biologist Brendon Larson argues that since we cannot reasonably hope for science to provide a wholly objective and literally true description of the world, we had best aim for an account that maximizes our chances of survival.

Objective science is still crucial, but for the practical objective of steering us onto a better path, metaphors may provide helpful visions of the goal and serve as effective roadmaps for getting there. Ecologists working in the fields of wildlife management and conservation biology are familiar with the tension that arises from trying to do properly objective science but for especially human ends, which means doing research strongly guided and informed by values that arise not from objective nature but from human society. A metaphor that straddles the science–society border (not that science can ever be fully separated from society) like "ecosystem health" may lead some to ask, "How can we talk objectively about such a thing when clearly it relies upon subjective human judgments about what is health and lack of it for an ecosystem?" Even more contentious is the metaphor that the earth as a whole is alive, a superorganism that attempts to regulate its various biological and chemical systems so as to maintain conditions suitable for life, a thesis that the environmental scientist James Lovelock called the Gaia hypothesis. Is there any way to make sense of these metaphorical terms and descriptions that is consistent with what we think of as objective science or is this just romantic poetry disguised in a white lab coat?

Ecosystem Health

Imagine being told by your local IT person that the reason your computer is not working is that it has become infected with a virus. Would your response be to say, "Wait a minute! Computers can't get sick with viruses, that's nothing but metaphor. You're supposed to be a computer *scientist* and I demand an objective literally true explanation!"? Probably not. Can we use this analogy to help us make sense of how it may be appropriate for ecologists and environmental scientists to use explicitly metaphorical and value-laden language to help us understand and solve environmental problems?

When the organismal principle of homeostasis is extended to an ecosystem or nature as a whole, it can promote talk of levels above the organism being healthy or sick. The metaphor of ecosystem or ecological health had been around for some time but really became popular within discussions of environmental monitoring and management circles in the 1990s. It has, however, proven to be a contentious concept. Is a healthy ecosystem one that has had no impact from humans? (Are there any ecosystems that have experienced zero human impact?) How much human impact and of what sort is consistent with a healthy ecosystem? What properties are relevant to assessing ecosystem health, and how does one go about measuring or counting them? Talk of "invasive" species suggests that the introduction or migration of new species into a region are bad for the local ecosystem or otherwise undesirable. But from the perspective of "objective" non-human nature, why or how could nature itself care about what is and has forever been an entirely natural process?

The problem is not that ecosystem health is a metaphor, but that it is such a vague and subjective one, open to quite different interpretations. As one critic states, "*Ecological health* is a nebulous concept that should be expunged from the vocabulary . . . There is no good objective and operational definition of what constitutes a healthy ecosystem, and searches for such a definition are a fruitless waste of time." Many critics note that ecosystem health is in the eye of the beholder. Others complain that not only is it anthropocentric, judging what constitutes health for nature from a human perspective, but that it is also far too subjective even for human purposes as people with different values may reach different conclusions about what a healthy ecosystem looks like. Because of its strongly anthropocentric perspective, many worry that decision-making on the basis of ecosystem health could legitimize the continued degradation of the environment for short-term economic gain.

At the root of the difficulty is that the metaphor assumes a suitable analogy between an individual organism (of which judgments as to health or disease are unproblematic) and an ecosystem. But an ecosystem is a poorly delimited, fuzzy sort of entity without clear spatial or temporal boundaries, and consequently application of attributes appropriate to organisms is problematic. Some of the features that have been considered as indicators of an ecosystem's health include: biodiversity, species abundance, productivity, integrity, and resilience. The last two are reminiscent of the contentious notion of the natural balance. But aside from that issue, there are questions about how to use any or all of these different value-laden measures to set public policy. What is the benchmark against which assessments of health or lack thereof will be measured? There seems often to be an assumption that a "natural" state unaltered by humans is healthier (i.e., "better" or "preferable" to one impacted by humans), but better or preferable for whom? For which species? How do we assess value and health at the level of a system as opposed to at the level of individual organisms? And do humans count as part of nature or not?

We do talk of a healthy work environment or a healthy society, but in doing so we really are evaluating the well-being of individual human organisms, not some higher-level entity; and to do so may be to commit what the philosopher Gilbert Ryle called a category mistake. This is to mistake something that properly belongs in one category of thing as if it belonged in another. For instance, if someone mistook talk about a horse's gait to refer to an anatomical part of the animal in addition to its legs, body, head, etc. rather than to the manner in which it moves, then they are committing a category mistake. It may be that all instances of metaphor are at least initially guilty of committing category mistakes. But as we've seen in other examples, sometimes what initially seems like a counterintuitive and literally false statement ("Organisms are made of cells") becomes an instance of polysemic speciation in meaning ("Organisms are not made of *prison* or *monk's* cells, they're made of *biological* cells"). So it should not be ruled out *a priori* that a similar shift in meaning could not occur with talk about the health of an ecosystem.

One thing that can be clearly said in its favour is that ecosystem health is a powerful rhetorical metaphor in public policy discussion, for who is likely to argue in favour of damaging the health of nature itself? But this does not address the pressing questions of how we are to decide whether preserving a swamp is "healthier" than draining it and backfilling it so that a park with a different (and possibly greater) diversity of plants and animals can be created. Critics urge that the term merely hides the value-driven assumptions and choices that should properly be open to public scrutiny, just as is done in other areas of public policy such as education, public health, and safety (compare the labels "pro-life" and "pro-choice"). For that reason, among others, some recommend that ecologists avoid the metaphor and simply lay out the specific options or proposals under consideration and their implication for specific species.

We have mentioned that the ecologist Frederic Clements considered an ecosystem as analogous to a living organism that goes through a series of developmental stages, reaching ultimately a "climax stage," characterized by a state of equilibrium from which the system, within limits, resists being moved. Clements referred to an ecosystem as a "superorganism," a term introduced by the entomologist William Morton Wheeler in 1911 to describe the similarities between an ant colony and a complex organism composed of many differentiated cells. (Recall from Chapter 5 the metaphors of the cell-state and references to the human body as a society of cells.) The geologist James Hutton spoke of the earth as consisting of three interlocked members: the "solid body of earth, an aqueous body of sea, and an elastic fluid of air," which together created a kind of living "physiology." While Hutton did suggest the analogy between the earth and a living organism, he also spoke of the planet and its interlocked processes as a well-designed "machine." Before the scientific revolution and the rise of the mechanical philosophy, the organicist philosophy was dominant, and the practice of construing the earth and the universe as a whole as analogous to a living organism can be traced back to Aristotle and even earlier in both Western and non-Western civilization. In the 1970s there was a resurgence of interest in organicist ideas through the promotion of the Gaia hypothesis by the chemist James Lovelock, who suggested that planet earth shared some features interestingly similar to a living organism.

Gaia

Lovelock had worked for NASA's mission looking for signs of life on Mars and other planets. Rather than attempting to identify living organisms directly, which in all likelihood will be microscopic, Lovelock decided it would be more profitable to look for indirect signs of life's presence, such as levels of atmospheric gases like oxygen and methane far from thermodynamic equilibrium, which would be consistent with the presence of living metabolisms but otherwise very improbable. The relatively stable levels of methane and oxygen in the upper atmosphere of the earth are improbable because methane naturally reacts with oxygen, eventually producing CO_2 in levels closer to thermodynamic equilibrium. That the levels have remained stable for millions of years is due to feedback systems resulting from the interaction of living organisms and abiotic geochemical processes. Lovelock called his attention to the processes that maintain the conditions suitable for life on earth "geophysiology," and it led to his hypothesis that living organisms actively regulate the composition of gases in the atmosphere in their own interests. The writer William Golding suggested to him that he call this idea Gaia, after the earth goddess of Greek mythology. Although it was not clear that Lovelock was intentionally suggesting that the planet was itself a living organism, or that the living things on it – again originally and still today by number and mass largely microbial – were knowingly or purposely engineering the planet to their benefit, the idea was taken up in popular circles as a validation of ancient beliefs and new-age philosophies associated with the burgeoning environmental movement. That probably did not help its chances of being accepted within the scientific community.

The idea was largely rejected by other scientists. Stephen J. Gould called it a metaphor in need of a mechanism. Richard Dawkins objected that since the planet cannot replicate itself to produce daughter planets with their own living systems and there evidently is no population of living planets on which natural selection could act, it does not qualify as an organism in any sense compatible with accepted evolutionary biology. And W. Ford Doolittle objected that while a global homeostatic regulation of oxygen and methane levels could occur as a result of biological activity, it is likely a fortuitous though improbable by-product of metabolic activity since it could not plausibly be encoded in the genomes of organisms so that it could evolve by natural selection. Others responded that if the content of the idea is simply that "Organisms and their environment evolve as a single, self-regulating system," that this has been well accepted in ecological science for some time, but is far from the more headline-capturing statements that the planet is akin to a living organism.

Lynn Margulis, the microbiologist and advocate of the endosymbiosis theory discussed in Chapter 6, helped to develop the Gaia hypothesis by supplementing evidence of further regulatory effects of algae and other microorganisms on global climate through biogeochemical feedback systems. This resulted in a revised version of the hypothesis maintaining that the planet comprises a complex coupled system involving the biosphere, atmosphere, hydrosphere, and lithosphere, and that "organisms and their material environment evolve as a single, coupled system, from which emerges the sustained self-regulation of climate and chemistry at a habitable state for whatever is the current biota." Natural selection, it is posited, would favour living organisms that contribute positively to the persistence of the local favourable environment, and this arrangement would gradually expand until there is a global cybernetic control system by which life is involved in regulating the conditions required for its own persistence, in analogy with a thermostat system that maintains room temperature in a house.

To answer the objections of Doolittle and other critics, Lovelock and Andrew Watson developed a model to show that such a feedback system could arise by natural selection. Daisyworld is a computer model consisting of two varieties of daisy, one white, the other black, growing on a simulated planet subject to increasing temperature like the earth, which has been exposed to increasing energy output by the sun as it has aged. (It is important to note that over much shorter periods of time the sun exhibits regular cycles of greater and lesser output, and the record-breaking temperatures on earth of the last two decades have occurred during a period of decreased solar output, so the current global warming cannot be ascribed to the sun.) White daisies would reflect sunlight enough from the planet to prevent it getting too hot; but if too many of the white variety grew, the planet would begin to cool due to their combined albedo effect, creating conditions more favourable to the dark variety, which would be able to absorb more heat and grow. But too many dark daisies would lead to global warming, which would encourage differential growth of white daisies able to reflect excess sunlight and so on, the result being a self-regulating climate system, and more importantly, Lovelock insisted, one capable in principle at least of developing through natural selection.

While not persuaded by the Daisyworld simulation, Ford Doolittle has recently begun to believe there may be a way to "Darwinize Gaia." This involves an extended evolutionary approach he has been developing that he refers to by the slogan "It's The Song Not The Singer." According to the ITSNTS perspective (or "It's nuts" as he humorously refers to it), one can allow for the selection of processes, in addition to entities like organisms or alleles. Doolittle uses the analogy that a song can persist even though the singers who sing it may eventually die or change from one performance to another. And if we think of life as a collection of interconnected biogeochemical cycles (like a choral concert) that persists even if the individual organisms and species (the singers) involved change over time or go extinct, we might thus "Darwinize Gaia."

The suggestion here is not that the Earth is itself alive or a living organism, only that it exhibits processes that help to regulate and assist in the persistence of life, and that these homeostatic feedback systems could evolve in an extended sense of Darwinian evolution that allows for processes to be units of selection that are subject to differential *persistence or survival* rather than differential *replication or reproduction*, as is the case with genes or organisms respectively in classical Darwinism. Biogeochemical processes whose properties promote their own persistence are naturally favoured by such selection; key among them is the ability to evolutionarily "recruit" (i.e., encourage the evolution of) diverse taxa that implement them. Notably, several of the steps of important biogeochemical cycles are catalyzed by the products of genes readily transferrable between taxa, thereby increasing the diversity of such "recruits."

The approach taken by Lovelock, Margulis, Doolittle, and others is to use metaphors heuristically to generate new testable hypotheses and perspectives that will generate novel questions and understanding of the natural world. It takes the idea of niche construction to a global scale and attempts to render it consistent with Darwinian evolution by means of natural selection.

What about the more popular accounts of Gaia that take the planet to be a living organism? Are there any persuasive reasons to take this metaphor literally? Here I think it pays to make a distinction between metaphors intended for scientific purposes (i.e., to generate empirically testable hypotheses and to provide empirically adequate explanations of natural phenomena that actually work in the sense of increasing the predictability of events and our ability to control their outcome), and metaphors whose intended effect is to change our own behaviour. If the point of Gaia talk is to make a factually true statement (that the earth is a living organism), then it arguably fails. But if the point is to frame our behaviour and its consequences in a way that will steer us onto a more sustainable path, then it may just be an effective rhetorical form of communication.

Many people worry whether the planet – or more accurately the biosphere – can withstand us humans. Are we perhaps the ultimate invasive species? Or are we, as many cynics like to say, more analogous to a cancer on the planet? Are we as a species singing the right song to ensure our continued presence on the planet or is it time to find another tune? We have the ability to affect the answer to that question. Practically speaking, if adopting the language of Gaia helps us to survive and thrive on the planet, it is beside the point to criticize it for being literally false (i.e., a metaphor). According to the tradition of many indigenous peoples of North America, chiefs are to make decisions with the interests of seven generations into the future in mind. These future generations of people, according to the Great Law of Peace of the Iroquois Confederacy, look up at the current generation from the ground beneath their feet. Literally speaking from the perspective of Western culture, it is false that future generations are present in the soil beneath our feet, but it is undeniably an insightful metaphor expressing a deep wisdom that, if broadly followed, could promote a much more environmentally sustainable way of life.

The trick, as always, is to balance respect for the different paths to wisdom that the many world philosophies and religions have created with a critical appreciation for the scientific values of objectivity, factuality, and truth. While we may be free to choose our own path forward, we should be foolish to discard the guidance afforded by the objective compass-reading that science provides. As many philosophers today insist, facts and values are intertwined; and while a compass can tell us how to proceed in a straight line, it cannot tell us in which direction we should want to go. The metaphors we adopt both reveal and influence the values that inform our decisions about where and how we choose to proceed. Science is not simply holding a mirror up to nature that reflects the objective truth; it is a tool, a methodology, and a way of thinking that increases our ability to understand and to survive in the world unequalled by wishful thinking or blind allegiance to cherished opinion. But to be used well it must be used wisely.

The next and final chapter is devoted to a set of influential metaphors in biomedicine which further highlight the performative effect of metaphors to prescribe certain avenues of research and interventions. As we increasingly use science to read the lines in the genomic book of life, we also find ourselves compelled to edit the text and our own futures, for better or worse.

8 Biomedicine

Genetic Engineering, Genome Editing, and Cell Reprogramming

Medicine, as the old saying goes, is as much an art as a science. Its chief objective is the treatment and prevention of illness and disease. Biomedicine combines the traditional and practical objectives of medicine with modern scientific understanding of normal and pathological function in humans and other living organisms, but especially from the perspectives of molecular and cellular biology.

As earlier chapters explained, the experimental and molecular turn in biology of the twentieth century involved the adoption of an engineering perspective in two senses: (1) the goal is not simply to achieve an understanding of how living things function through passive observation, but rather to determine by intervention and manipulation of their component parts (cells and their own components) how they behave in different environments and conditions; and (2) this approach is guided by engineering metaphors that portray organisms and cells as machines that can be disassembled and rearranged in order to learn how the parts and whole function. The human body, in other words, is conceived as a complicated machine or system of integrated mechanisms, and pathology or absence of normal ("good") health is the result of some malfunction of the components of one or more of these mechanisms or the interaction among them. The practical goal of biomedicine, therefore, is to "repair" the malfunctioning part(s).

Much could be said about the practice of modern medicine and the delivery of healthcare, and how each is framed by common metaphors of our time, how cancer is described as a *battle* that must be *courageously fought or lost,* or how speaking of the COVID-19 pandemic in terms of *waves, wildfires,* and *storms* may create feelings of helplessness and undermine adoption of necessary public health measures. But the discussion of biomedicine here will be more focused on how metaphor influences the scientific understanding of living cells and organisms and attempts to use this knowledge to treat and prevent illness and disease through biomolecular intervention. The molecular revolution that drives the modern biomedical approach has made possible biotechnology, gene therapy, and synthetic biology. Consequently, this chapter focuses on the key metaphors of *genetic engineering, genome editing,* and *cell reprogramming* as they are relevant to human health and medicine. Note that these metaphors all involve action verbs: engineering, editing, and reprogramming in contrast with the noun metaphors of books, blueprints, and programs discussed earlier. What characterizes genetic or genomic engineering and synthetic biology is a focus on building things with and from organic matter for commercial and health-related reasons.

As they appear in a medical context, these metaphors project an attitude of confidence that we can use our scientific understanding of how the body works to manage and possibly cure many diseases. They speak of a promise that we will take control of our health destinies through rational intervention and reconfiguration of our genomic and cellular makeup. But the potential to use the tools of molecular biology to enhance and to engineer human embryos and adults also raises unique ethical challenges and fears of a renewed and more powerful version of eugenics. The literature dealing with the moral and societal dimensions of genetic and genomic engineering is extensive. But this chapter (like the rest of the book) will concentrate specifically on the metaphors with which these possibilities are articulated and the issues framed. What work are these metaphors of genomic engineering, editing, and reprogramming doing exactly? What features and questions do they foreground? Which do they push to the margins of our attention or block altogether? Are they the best metaphors available? And what are the alternatives?

Because the biomedical objectives to be discussed in this chapter are as practical, if not more so, than they are theoretical (in the interest of so-called pure science), it may be appropriate to regard the metaphors with which they are articulated as serving especially instrumental functions, as *tools* rather than as pictures or descriptions intended to objectively mirror reality. If the point is to make someone well, to alleviate pain, (or perhaps to enhance the cognitive or physical abilities of a future human being), one might take the position that asking whether our cells or genomes are *really* machines or computers may be less to the point than asking whether thinking about them in these terms helps to achieve the desired results. Still, even if one is comfortable with this instrumentalist attitude, it is legitimate to ask whether these metaphors (and what the philosopher of science Dan Nicholson calls the Machine-Conception of the Organism with which they are closely tied) are likely to help achieve the practical objectives for which they are employed, or whether they might mislead us into a future we would rather avoid.

Genetic Engineering

Genetic engineering refers to the manipulation of an organism's genome by use of recombinant DNA technology. When scientists working in a lab isolate DNA sequences from the cells of one organism and insert them into the genome of another, the two DNA sequences are said to have been recombined. The first such laboratory recombination of DNA from separate organisms was achieved by Paul Berg and his team in 1972, who "spliced" together DNA from two separate viruses, the monkey virus SV40 and lambda virus. Berg had concerns about the potential risks of creating novel forms of virulent pathogens and so did not proceed beyond recombining genetic material in a lab dish (*in vitro*). But the following year Herbert Boyer and Stanley Cohen adapted Berg's technique to create the first genetically modified organism by combining DNA from two separate strains of the bacterium *E. coli*. Very shortly after, they introduced genes from the frog *Xenopus laevis* into *E. coli* that was successfully transcribed into RNA.

The process is commonly described as involving the "cutting" of DNA with a restriction enzyme from one source and "pasting" it with a ligase enzyme into another. The transplanted DNA is then said to be "cloned" by the host cells, as the latter replicate the newly recombined genome as they divide. The result of the transfer of DNA from one distinct species to another is also referred to as "chimeric" DNA, after the mythical beast combining features of several distinct animals. Recombinant DNA sequences can alternatively be replicated or "amplified" in a test tube using polymerase chain reaction (PCR). By these means, novel genes can be inserted or alternatively "knocked out" or "silenced." These techniques of genetic engineering are responsible for the biotech industry that has revolutionized the production of pharmaceuticals (it was Boyer who turned bacteria and yeast into literal cell factories by inserting into them the human insulin gene), new forms of genetically modified agricultural crops and animals (from tobacco, potato, canola, soy, corn, and rice to salmon), and model organisms for the laboratory study of diseases such as cancer (e.g., the Harvard OncoMouse).

Genetically modified organisms (GMOs) created in government, university, or private laboratories have been approved for patent protection in many countries. As a consequence, many critics object that the practice reflects the pernicious influence of materialism and is a radical step in the further commercialization of life by which whole lineages of living things (not just specific individual plants or animals) have been turned into private property. Those who believe life to be the sacred creation of a supernatural being are understandably opposed to this development, but many secular people also have concerns about the safety of GMOs, and about the political, economic, and social implications of the further concentration of control over food and health resources by private profit-seeking interests in a competitive capitalist system.

Traditional gene manipulation techniques were comparatively crude and dependent on trial-and-error "shotgun" approaches that involved either inserting the foreign DNA by means of a viral or bacterial vector, injection via micropipette, or by a delivery system known as a "gene gun." The results would then need to be screened to determine whether the intended DNA sequence had been successfully integrated into the host genome, and if possible where in the genome the insertion had occurred. In addition to being able to "read" the genetic code of DNA, scientists have also learned how to design and synthesize specific DNA sequences from its nucleotide "building blocks," which has led to the common talk of scientists also "writing" in the language of DNA.

The term "genetic engineering," then, dates back to the 1970s, but its use really took off in the late 1990s (according to the PubMed database of biomedical and life science literature). Talk of gene therapy also grew in the 1990s and occurrences of "molecular medicine" have sharply increased in the last 10 years. "Engineered cell therapy" is a new term just gaining attention in the last several years. But one of the most significant developments recently in the discourse of genetic engineering is the effort to "edit" the genes or genomes of humans and other organisms.

Genome Editing

This is perhaps a natural extension of the talk of scientists' ability to "read and write" in the "language" of DNA, a language that has long been conceived as a computer code in which the programs for cell and organismal function and development are written. Given this background of computer engineering, it is hardly surprising that scientists now seek to edit the "programs" and "circuits" believed to underlie many diseases and disorders. There are scattered occurrences of the terms gene or genome editing in the 1980s and 1990s, but a much-cited review of the field in 2010 by Fyodor Urnov and collaborators at the biotech firm Sangamo BioSciences demonstrates that the original term of preference was "gene targeting." The first efficient techniques for making targeted modifications in DNA employed endonucleases (restriction enzymes that make cuts to the sugar–phosphate backbone of DNA) bound to DNA binding transcription factors known as Zinc-finger proteins in the mid-1990s and TALENS (transcription activator-like effector nucleases) beginning in 2010. (The term "Zinc-finger" derives from the finger-like protrusions of molecular domains in these proteins, some of which bind to molecules of zinc.) Both techniques rely on the sequence-specific binding ability of transcription factors to deliver the DNA-cleaving nucleases to their selected target. These techniques have been used to "disrupt," "silence," or "knock out" genes of choice to study their effects in the development and function of cells and whole organisms; and by supplementing the DNA-cleaving nuclease with a nucleotide sequence of their design, researchers can rely on the cell's natural DNA repair system to "correct" or "add" genes to the targeted genome.

To highlight the greater ease and specificity with which specific loci could be targeted using these methods, Urnov and his colleagues recommended adopting the terminological innovation of "genome editing." While these techniques did allow scientists to make reasonably reliable and specific alterations to eukaryote genomes, including those of human cells, they were expensive and difficult to use successfully. The recent development of CRISPR/Cas9 as a new gene editing technology has made the "cutting-and-pasting" of genes cheap and easy in comparison and has quickly become the tool of choice in labs the world over.

CRISPR stands for "clustered regular interspaced short palindromic repeats," and refers to nucleotide sequences first observed in 1993 in bacteria and archaea. These DNA patterns consist of a sequence followed by nearly the same sequence in reverse, then a "spacer" sequence of 30 or so seemingly random bases, a repeat of the first sequence and its palindrome, then a different spacer sequence, and so on. At first the function of these repetitive DNA sequences was unknown, until a similarity was noted in 2005 by Francisco Mojica between the spacers and viral DNA. By 2007 evidence confirmed Mojica's suggestion that CRISPR and the family of nucleases whose genes are closely associated with them evolved in bacteria and archaea as a form of adaptive immune system response to infection by viruses (bacteriophages). The spacer sequences from the viral DNA are taken up in the bacterial genome (horizontal transfer again!) as a kind of memory bank of past viral infections. Together with the nucleases that are associated with them (Cas for *CRISPR-associated*) they form a complex macromolecular machine that, in the event of a future reinfection, binds with the assistance of a complementary RNA guide to a specific site in the viral DNA, at which point the associated nuclease protein cuts and disables the foreign DNA.

It wasn't long before a team of scientists led by Jennifer Doudna and Emmanuelle Charpentier recognized that by designing RNA guides to recognize specific DNA sites of interest, the CRISPR/Cas system could be engineered to make targeted cuts in the genome of any organism, including humans, to potentially eliminate a disease-causing gene mutation. This occurs as a result of the cell's natural DNA repair system, which in attempting to splice the broken strands together may create a mutation that either makes nonsense of the gene coding sequence or renders it less effective at synthesizing its associated protein or carrying out its gene regulatory function. Alternatively, by coupling a normally functioning gene with the CRISPR/Cas9 system, the cell can be coaxed to replace one base for another (an A for a C, for instance) or an entire faulty gene with a normally functioning one.

The Doudna–Charpentier team published in 2012 the results of their work on the bacterium *Streptococcus pyogenes*, showing that the CRISPR/Cas9 system could be "programmed" to make "edits" at specific DNA loci. They wrote:

> Our study further demonstrates that the Cas9 endonuclease family can be programmed with single RNA molecules to cleave specific DNA sites, thereby raising the exciting possibility of developing a simple and versatile RNA-directed system to generate ds [double-strand] DNA breaks for genome targeting and editing.

Other labs led by Feng Zhang and George Church soon published results showing that they had engineered the CRISPR/Cas9 system to make edits in human cell culture lines. A contentious (and ongoing) legal battle ensued over which team deserved priority for patent protection of the CRISPR system as a technology for genome editing. Its ease of use and low cost has led to what is called the "CRISPR revolution" in biology and biomedicine, a claim amply supported by the announcement in October 2020 that Charpentier and Doudna have been awarded the Nobel Prize in chemistry "for the development of a method for genome editing."

Various CRISPR and Cas nuclease systems (also commonly called "platforms") are now widely deployed to "silence" or "knock out," to "cut-and-paste," or to "amplify" the transcription of desired gene "targets" in a wide range of organisms and cell lines. Cas endonucleases (including helicases that "unzip" the double-stranded DNA and nucleases that "cut or cleave" DNA) are "guided" by small RNA molecules designed to "recognize" specific sites in the "target DNA."

CRISPR as Programmable Molecular Scissors

Nucleases that break, cleave, or cut DNA have for some time been described metaphorically as molecular "scissors," and now in the context of gene editing these Cas nucleases are commonly associated with the "cut-and-paste" editing function of word-processing software (Figure 8.1).

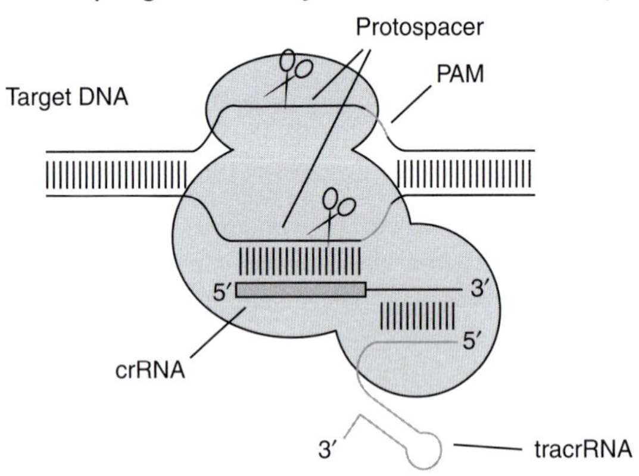

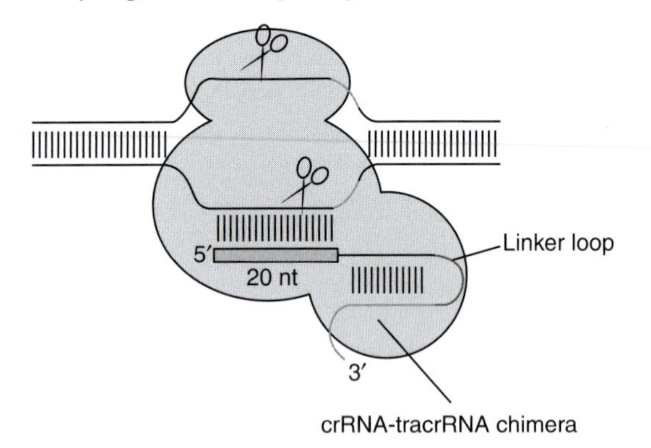

Figure 8.1 CRISPR and associated endonuclease (Cas) portrayed as molecular scissors. (redrawn and adapted from Jinek, M., Chylinski, K., Fonfara, I., et al. (2012). A programmable dual-RNA-guided DNA endonuclease in adaptive bacterial immunity. Science 337: 816–820, fig. 5a).

In a widely viewed TED Talk of 2015 Doudna explained, "We can think of older genome engineering technologies as similar to having to rewire your computer each time you want to run a new piece of software, whereas the CRISPR technology is like software for the genome, we can program it easily, using these little bits of RNA." Elsewhere she has written that CRISPR promises to make the genome as "malleable as a piece of literary prose at the mercy of the editor's red pen." Without detracting from the significance of the scientific achievement, it is probably fair to characterize this language as serving a largely rhetorical function. Both the editing and software metaphors are likely also examples of what Dorothy Nelkin called "promotional metaphors," whose purpose is to attract attention to a scientific achievement or topic and to create confidence in government funding agencies and potential venture capital investors from whom the speaker is hoping to secure financial support. Doudna and others involved in the development of CRISPR and other gene editing technologies have created biotech companies through which they hope to capitalize on commercial applications of the science.

The use of this gene editing technology on humans has, however, come under a good deal of scrutiny. Doudna herself and other scientists have called for a moratorium on its application to humans until its safety can be assured. One source of concern is the potential for what are known as "off-target" effects occurring at other DNA sites in the same cell or in other unintended cells and organs of the body. As we saw in Chapter 3, genes provide some (but not all) of the essential instructions for building various types of proteins, but the same protein may be expressed in many different cell types in distinct tissues and organs, where they may even be involved in quite different functions and physiological pathways. So altering a gene with the intention of bringing about one effect in one part of the body may actually end up having unintended consequences in multiple other sites. And this is a possibility even assuming that CRISPR is a perfectly precise and predictable gene editing tool that only targets the DNA sequences intended, which unfortunately as of yet it is not. There are additional concerns that the technique can itself unintentionally cause DNA mutations and chromosomal damage, or even loss of an entire chromosome.

Such concerns, however, did not stop the scientist He Jiankui from using CRISPR/Cas9 in 2018 to genetically edit the genomes of twin girl embryos created by *in vitro* fertilization (IVF) to protect them from infection by the virus linked to AIDS. The embryos were subsequently implanted into a woman (whose husband is HIV-positive) and successfully born, evidently so far without any complications. (Critics noted that risk of infection from an HIV-positive male can be nearly eliminated through IVF and a procedure called "sperm washing" to remove any viral particles from the semen, and that the twins are not immune from infection after birth because the virus can enter cells through another cell membrane receptor not disabled by Jiankui's experiment.) The affair illustrated the especially sensitive issue of editing human germline cells, which in contrast to manipulating the somatic cells of an individual person could be passed on – along with any harmful consequences – to future generations of people derived from those cells. After universal condemnation by the scientific and ethical community, Jiankui was sentenced by the Chinese government to three years in prison for violating his university's and national laws against such human experimentation.

In 2019 a patient with sickle-cell disease received a transfusion of CRISPR-modified red blood cells and in 2020 another person with a rare inherited eye disease (Leber's congenital amaurosis 10) was treated with CRISPR-modified virus cells that were injected directly into the retina with the objective of deleting the mutated gene responsible for the condition. (One of the biotech companies involved in this latter experimental therapy is called Editas Medicine.) These interventions were first attempted on human cell cultures, mice, and monkeys to show they are safe and effective; and they do offer great hope to many people who have no alternative. It is still early, but so far the intervention appears to have worked very well for at least the first patient. Earlier trials of gene editing have been performed in humans using Zinc-finger techniques.

But as Elinor Hortle, a medical researcher who uses CRISPR/Cas9 in her own lab, has noted, the metaphors of molecular scissors and genome editing are at this stage of development misleading because they suggest a far greater degree of accuracy, effectiveness, reliability, and safety than is yet warranted for use on humans. Hortle suggests replacing the simplistic genome editing and cut-and-paste imagery of what is involved in using CRISPR to modify the behaviour of a cell with the much more complex and uncertain analogy of attempting to prevent soccer hooligans from rioting in a city by manipulating a bit of the online code for the FIFA rule book. In her analogy, CRISPR is a bit of malware that can identify specific 20-character phrases of HTML code.

Even assuming you can make precise changes to the code without difficulty, you have far less confidence that you will change only the 20-character phrase that occurs in the FIFA rule book and in no other site on the Internet, or that soccer fans will read the edited version, or if they do how they will respond in any particular city after the outcome of any particular match. By analogy, there is a series of complex steps that must take place between the modification of DNA in a cell and the ultimate phenotypic or observable result. A brief and highly abridged list of these steps includes: transcription into pre-messenger RNA, removal of intron segments and splicing to create an mRNA, translation into a polypeptide chain, folding into a three-dimensional protein conformation, further post-translation modifications, and subsequent protein–protein interactions – and of course none of these processes occur in a vacuum free of external environmental influences.

Moreover, as Hortle explains, "by thinking the problem this way, we've just given ourselves a pretty decent feel for the complications of polygenic disease, incomplete penetrance, missense/nonsense mutations, epigenetic silencing, genetic compensation, off-target and germline effects – all without a single word of scientific jargon." Her malware analogy is not likely to supplant the molecular scissors account (nor does she suggest it will), but does provide an informative illustration for non-scientists of how complicated the efforts to perform molecular medicine/gene therapy is. And as Brendon Larson has argued, ensuring that there are alternative metaphors in play not only creates what he calls a "fruitful friction" that keeps the scientific imagination alive, but it is also a prophylaxis against any single metaphor and perspective dominating both the scientific and popular understanding of potentially important societal implications of science and technology.

The historian of biology Michel Morange notes that the choice of the term "editing" is a natural extension of the metaphorical description of the human genome as the "book of life," but that it also "suggests to the public a nonthreatening, merely technical approach to gene manipulation." Many metaphor scholars have asked whether the metaphors used to describe CRISPR/Cas9 and other gene-modification techniques provide an accurate assessment of the intervention's safety and reliability. Metaphors of editing, cutting-and-pasting, and molecular scissors all project an image of highly precise and innocuous activities that only lead to improvement. But if we do choose to think more carefully about genome editing as analogous to the find-and-replace function of word-processing software, should we not consider the analogous possibilities that intentional changes to one word/nucleotide sequence may have unintended effects on similar sequences (for instance, might changing "cat" to "cap" prove to have "capastrophic" consequences?); that edits made in one document might unexpectedly also turn up in other documents (cells and organs); or that editing the germline raises the possibility that every future document you write or download to your computer may be similarly edited, whether you desire it or not. In fact, speaking in Mary Hesse's terms, these are all known to be positive analogies, to varying degrees and dependent on the type of cell, organ, and organism on which the edits are being made.

Description of the Cas9 nuclease as a molecular "scalpel," by contrast at least, suggests a need for surgical skill and the presence of risk and uncertain results, and indeed the use of CRISPR for the medical treatment of humans is now frequently referred to as "genome surgery." Similarly, Meaghan O'Keefe and her colleagues note that talk of using CRISPR to "target" genetic defects is preferable to the editing metaphor, because it suggests that there is an element of risk involved, especially in highlighting the possibility of "off-target effects."

How essential are these metaphors for the science itself, as opposed to serving largely rhetorical roles in communicating with the non-scientific public? The scissors, cut-and-paste, and scalpel metaphors are not quite theory-constitutive, I would argue, because other terms can be and are used to describe the activity of the Cas nucleases (e.g., cleaving, breaking). They are, however, effective means of communicating to the public in simple terms what is in reality very complex molecular biology. It also seems clear that they serve an additional function as promotional metaphors intended to inspire confidence in the medical and bioengineering projects in which they are playing an increasingly important role.

There is a better case to be made that the editing metaphor is theory-constitutive, and therefore influences how the scientists themselves understand their research objectives. The metaphor of gene and genome editing is a rather natural outgrowth of the earlier root metaphors that DNA is a code, a blueprint, an instruction book, and a program for protein synthesis and cell and organismal development. To the extent that these earlier textual metaphors have become integral to how many scientists understand genomes and development (for better or for worse), it could be argued that thinking about the efforts to intervene in the molecular genetics and biology of humans and other organisms for health and other purposes in terms of editing has likewise become part of the very scientific theory and more than a rhetorical flourish by which scientists communicate with the public. For that very reason, greater caution and self-awareness that genome editing *is* a metaphor is all the more important.

But it is clear that many scientists actively engaged in CRISPR research are aware of the difficulties attending these metaphors. For example, Fyodor Urnov, currently the scientific director at the Innovative Genomics Institute, used a sobering analogy to describe the occurrence of significant off-target deletions and chromosomal rearrangements in the experimental use of CRISPR/Cas9 to correct genetic defects in non-viable human embryos: "If human embryo editing for reproductive purposes or germline editing were space flight," he recently wrote, "the new data are the equivalent of having the rocket explode at the launch pad before take-off."

Urnov advocates for the use of somatic cell gene editing for therapeutic purposes like the treatment of patients with sickle-cell disease or hereditary blindness, but is staunchly opposed to its application to human germline cells or embryos. For one, there is no chance for informed consent in these cases, and any unintended damage would be irreversibly passed on to each cell in the body and to future generations should these individuals grow up to have children of their own. This is why He Jiankui's experiment on the twin embryos is so alarming and controversial.

New research using a family of "programmable" enzymes known as adenosine deaminases to edit mRNA would provide an alternative form of molecular therapy that, unlike CRISPR and other genome editing techniques, would be non-permanent and less likely to have off-target effects. In addition, editing a person's own mRNA decreases the chance of provoking an immune response, which is a possibility with CRISPR therapies that use the bacterial-derived Cas9 nuclease. And because mRNA is quickly degraded by the cell and is not part of the permanent DNA "blueprint" or "program," edits made to it will not be replicated when the cell divides, nor will it be passed on to future generations.

While we speak of editing a text or computer code, we say that one engineers a computer or machine. Consequently, we see widespread talk of combating disease at the cellular and molecular level in terms such as reprogramming cells and rewiring genetic "circuits" (although this too is done in many cases by making edits to the genome). Drug design involves identifying molecular targets in nucleic acids (DNA or RNA) or proteins to which a chemical compound can bind so as to influence the circuits or pathways in which they are implicated. These "circuits" and "pathways," it should be noted, are not fixed physical entities as the metaphors suggest, but refer rather to diagrams and other visual aids used to represent spatially in two dimensions the temporal organization and sequence of interactions that are believed to typically occur among an often very large set of molecules, including extracellular signals, membrane-bound receptors or channels, intracellular second messengers, kinase "switches," transcription factors, DNA sites, or the "tails" of histone proteins that make up the nucleosome units around which DNA is wound to form the chromatin material of which chromosomes are composed. They are closer to organizational charts for a large institution (showing who reports to who) than they are to electronic circuit boards or maps of a geographical terrain.

Notwithstanding the fact that cells are not literally machines or computers, it has been heuristically productive to think of the cell *as if* it were an engineered device composed of rationally intelligible mechanisms, similar in its organization and mode of operation to human-made machines. Regarding the cell as though it were a computer has helped scientists to break its complicated function down conceptually into distinct processes or "subroutines" and "programs"; and if they can successfully use these models (which we can think of as hypotheses) to better understand what is going on in the cell and the body as a whole, the hope is that they can then nudge it into behaving in ways more favourable to our health and well-being when it goes awry. Just how far these comparisons and analogies will ultimately prove useful is still unclear, though we have seen already in previous chapters the arguments for why they should be viewed with some suspicion. But what exactly *are* scientists talking about when they speak of "reprogramming" and "rewiring" cells?

Cell Reprogramming

The key background metaphors against which this research in synthetic biology and biomedicine is arranged is that the cell is a *computer,* DNA is a *computer code,* and the genome *encodes* a complex network of developmental and physiological *circuits* that guide the behaviour and interaction of a *cast of molecular characters* consisting of charged ions, gases, lipids, peptides, proteins, protein complexes, and nucleic acids. The mixed metaphor here is intentional, for despite the apparent primacy of the computer engineering metaphor, much of the description of the behaviour of the molecules involved in these circuits is in terms of actors and agents that interact, signal, recruit, compete, and cooperate with one another to carry out various functions within the cell (see Chapter 4).

Stem cell research is the major area in which talk of reprogramming cells occurs. "Stem" cells are those capable of developing or "growing" (like the stem of a plant) into distinct differentiated cell types. Embryonal stem cells (the small mass of cells that develop within the hollow early-stage blastula) are pluripotent, meaning they can develop into any of the body's differentiated somatic cells, while the cells of the outer lining of the blastula are totipotent, meaning they can also develop into the chorion and fetal placenta tissues that connect the embryo to the mother's womb throughout pregnancy. Differentiated somatic cells are said to be reprogrammed for "stemness" or for alternative tissue-specificity. The medical applications of this research include *in vitro* modelling of diseases, drug testing, assessment of gene and cell therapies, and the regeneration of damaged or missing tissue. *Cellular reprogramming* has its own journal (a key indicator that a research topic has truly come of age), edited by none other than Sir Ian Wilmut of "Dolly the sheep" fame. The journal was first launched in 1999 under the title *Cloning,* amended in 2002 to *Cloning and Stem Cells,* and changed again to its current title in 2010 "to reflect all mechanisms of cellular reprogramming."

To reprogram a cell means to change its *fate* – that is, the phenotypic characteristics and gene expression patterns that a cell's development concludes in. (So, in a strange way, it means to change a cell's fate after the fact.) As embryonal stem cells divide, the so-called daughter cells are shuttled down different "pathways" in a developmental "landscape" to become specialized somatic tissue cells, such as neurons, myocytes, fibroblasts, red or white blood cells, and egg or sperm cells. (The influential metaphor of an "epigenetic landscape" was introduced by the biologist Conrad Waddington.) More correctly, when a stem cell divides, one of the daughter cells begins to express genes specific to becoming a distinct type of differentiated tissue cell while the other remains in a less differentiated stem state, thus maintaining a stock of stem cells for regeneration and repair if needed. Beginning in the late nineteenth century, embryologists produced "cell fate maps" and "cell lineage trees" detailing the developmental history of tissues and organs from the early embryonal cells. "Niches" of multipotent and unipotent adult stem cells have been found in many of the body's tissues and organs with varying degrees of ability to differentiate into various cell types.

Reprogramming involves either reverting a mature, differentiated somatic cell into a more embryonal like state, resulting in what is known as an induced pluripotent stem cell (iPSC), or to a different somatic cell state (called transdifferentiation). Conceptually (and metaphorically) the adult cell is pushed backward as it were, uphill through the developmental landscape or over a hill into a neighbouring valley (Figure 8.2) by exposure to specific transcription factors or through the erasure and re-establishment of epigenetic marks (chemical "tagging" of histones and chromatin "remodelling") laid down during cell development. An earlier form of reprogramming by means of somatic cell nuclear

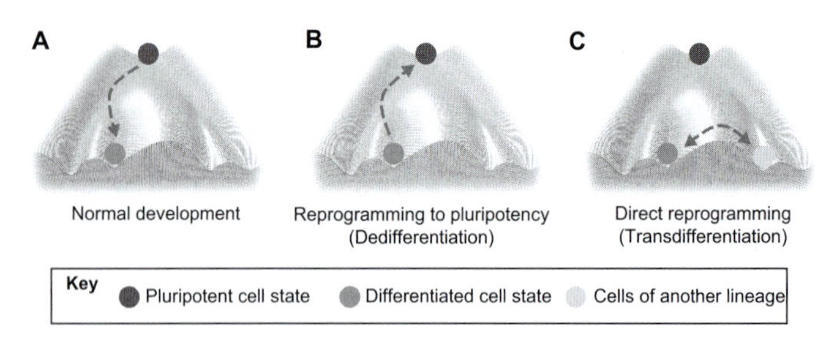

Figure 8.2 Reprogramming the epigenetic landscape (reproduced with permission from The Journal of Cell Science. From Takahashi, K. (2012). Cellular reprogramming: lowering gravity on Waddington's epigenetic landscape. Journal of Cell Science, 125 (11): 2553–2560, fig. 1. https://doi.org/10.1242/jcs.084822).

transfer (in which the nucleus from one cell is injected into another whose nucleus has been removed) was used to create cloned mammals like Dolly.

CRISPR/Cas9 is now also being used to create "programmable artificial transcription factors" to switch on or off endogenous genes in human cells. In these cases, the Cas9 scissors are deactivated (dCas) and equipped with a single-guide RNA designed to recognize and bind with a specific gene sequence. In CRISPRa, a gene is activated or "switched on" by associating the dCas enzyme with a transcriptional activator. CRISPRi is used to interfere with transcription, effectively switching the gene off. In both cases it is the epigenome (the collection of molecular tags and markers that attach to chromatin) being edited rather than the genome itself, as no cutting of the DNA is involved and no mutations, deletions, or additions of bases are involved.

Clearly it only makes sense to refer to these sorts of interventions into cell phenotype as *reprogramming* against the background metaphor that the cell is a computer. But it is by no means inevitable or necessary to adopt this language and way of thinking about cell therapy. Given the older convention of talking about cell fate, scientists could, for instance, think of their efforts as attempts to "retrain" or "re-educate" somatic cells. We already speak of immune cells as being trained and schooled to recognize foreign cells and viruses. But perhaps re-education sounds too much like a state-controlled propaganda program, and perhaps because these terms have been used in reference to immune cells, the metaphor does not appear available for extension to somatic cells generally.

However, it seems quite likely that the programming language is attractive to scientists because it provides an analogical *mechanism* by which they can understand more precisely how to achieve the results they are after. It's one thing to say you want to retrain or re-educate a cell, but how then do you go about it, precisely? Given the existing prevalence of the genome as computer code metaphor and scientists' ability to "read and write" in this code, it is a rather natural analogical extension to think of reprogramming cells to behave in desired ways. In fact, computer scientists and molecular biologists at Microsoft's research lab at the University of Cambridge are collaborating to "decode" the "bug" in the genetic "program" that they believe causes cells to begin the uncontrolled replication that results in the growth of a cancerous tumour. In fact, according to this particular model of cancer, a cell must suffer several genetic mutations, resulting in several faulty circuit switches, before it acquires the several "hallmarks" of malignant neoplasia. For this reason, "switching off" cancer is much more difficult than turning off an annoying blinking light or buzzing electronic device.

A closely related engineering metaphor used in relation to cell therapy is talk of "rewiring" cells. According to this approach, cell function is determined by a series of genetic and signalling "circuits" which, if impacted by genetic mutations or other injury, can result in abnormal and dysfunctional physiological behaviour that underlies disease at the cellular, tissue, organ, and organismal levels. What exactly is the relation between cell reprogramming and cell rewiring?

Rewiring Cell Circuits

Talk of rewiring cells can refer to changing cell fate by reprogramming, as discussed above, or to changing the physiological behaviour of a mature somatic cell in response to external signals. The "wires" in question refer to gene circuits (the network of genes and gene products that regulate gene transcription and protein synthesis) and/or signalling pathways (the network of extracellular signals, membrane receptors, and intracellular messengers and molecular effectors that govern cellular response to a signal). In reality, there are no literal wires within a cell; rather, the term refers to the conceptual organization of the functional relationships and interactions that typically occur between various cell components in specific temporal sequences. Genetic circuit and gene regulatory network models are extensions of the lac-operon model first developed by Jacob and Monod (discussed in Chapter 2), that provides an account of the causal relationships between gene promoter, repressor, and operator units that interact in positive and negative feedback loops to govern the expression of a gene in response to an extracellular signal. Molecular biologists use wiring diagrams to represent in a flat two-dimensional form the "logical architecture" of the relationships and interactions between various molecular components in the cell that activate and repress one another like the switches and transistors of an electronic circuit board (Figure 8.3).

Those working in synthetic biology look to apply the principles of genome editing and protein engineering (rearranging natural amino acid sequences

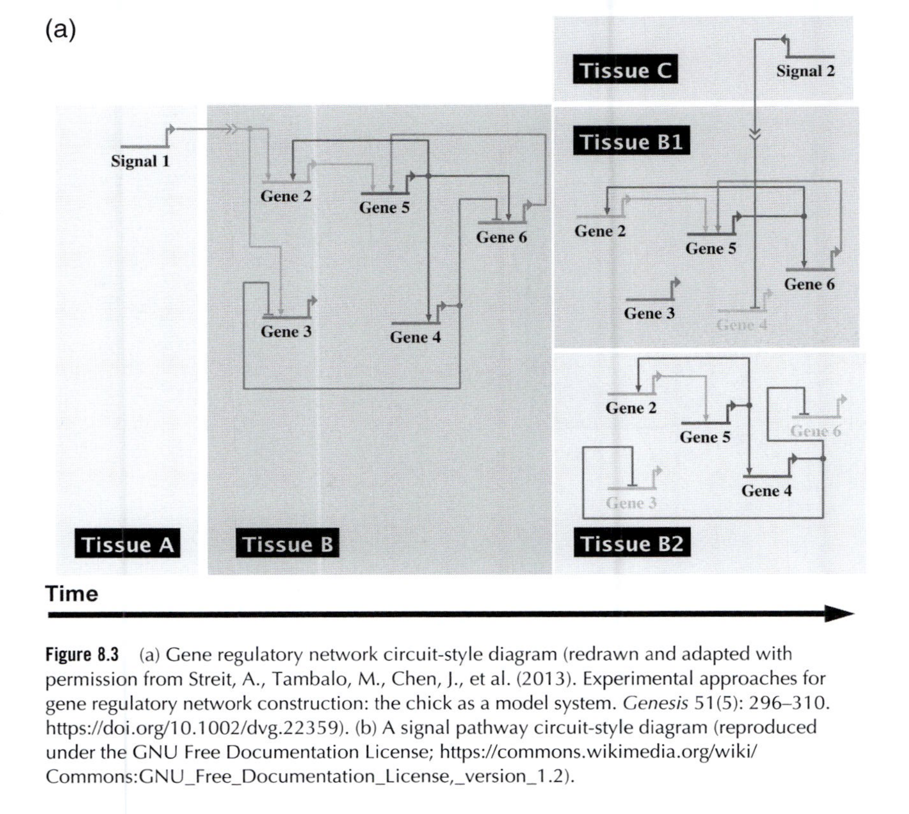

Figure 8.3 (a) Gene regulatory network circuit-style diagram (redrawn and adapted with permission from Streit, A., Tambalo, M., Chen, J., et al. (2013). Experimental approaches for gene regulatory network construction: the chick as a model system. *Genesis* 51(5): 296–310. https://doi.org/10.1002/dvg.22359). (b) A signal pathway circuit-style diagram (reproduced under the GNU Free Documentation License; https://commons.wikimedia.org/wiki/Commons:GNU_Free_Documentation_License,_version_1.2).

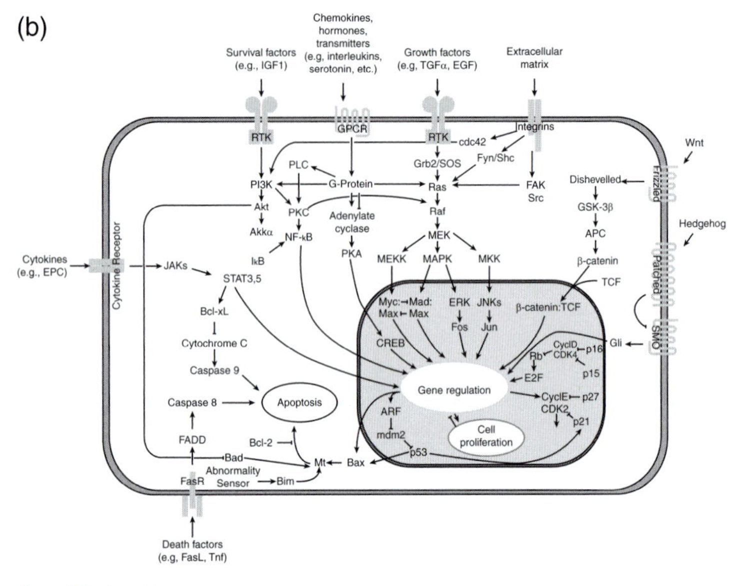

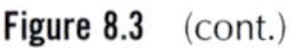
Figure 8.3 (cont.)

and designing novel ones) to redesign cell function and behaviour. One such application in biomedicine is the "rewiring" of T-cells (a type of immune cell) to launch an immune response when in the presence of cancer cells that secrete specific surface proteins that can suppress immune cell activity. These reprogrammed immune cells carry a laboratory-designed receptor (known as a Chimeric Antigen Receptor (CAR)) that is engineered to recognize and bind to a specific cancer cell surface protein. These CAR T-cells promise to launch a very specific immune response to selectively kill cancer cells rather than indiscriminately killing healthy cells as injection of lymphocytes or chemotherapy drugs into the patient's bloodstream do (Figure 8.4).

Another project uses a deactivated Cas9 enzyme to "rewire" the response of a line of immortalized human T-lymphocytes used to study lymphoma to a specific signal. Using a deactivated Cas9 enzyme, Martin Fussenegger's lab at the ETH Zürich have engineered multi-complex molecular units they call

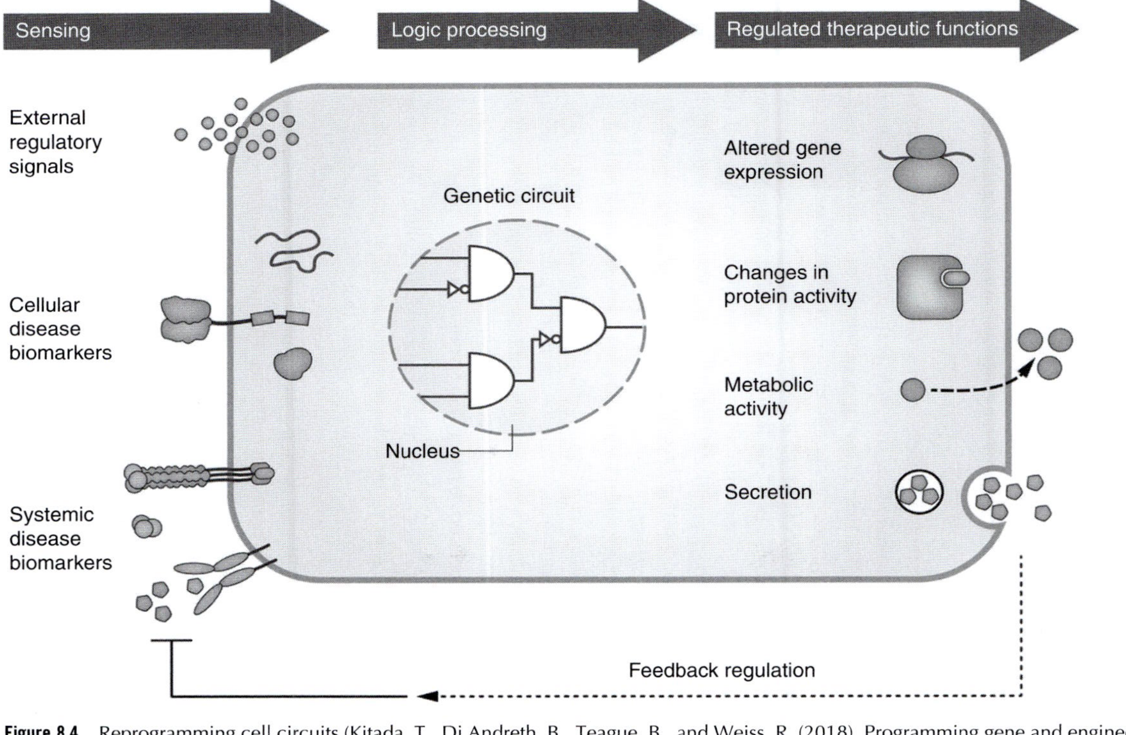

Figure 8.4 Reprogramming cell circuits (Kitada, T., Di Andreth, B., Teague, B., and Weiss, R. (2018). Programming gene and engineered cell therapies, Science 359(6376), fig. 1; reprinted with permission from AAAS). (Kitada et al. 2018).

GEARs (generalized engineered activation regulators) so that T-cells respond in novel ways to the reception of common external signals, thereby "rerouting" or "hijacking," in their words, several common signalling pathways in therapeutically positive ways. The invocation of "gear" language here may seem an incongruous mix of mechanical and computer technology, but it is consistent with the engineering language and imagery popular among synthetic biologists.

Although synthetic biology's application to biomedicine is proceeding at great speed, it is still uncertain how successfully these developments can be translated into clinical practice. But what does seem clear is that scientists are not inclined to stop using computer and engineering metaphors to conceptualize the future possibilities of this research anytime soon. Nor are they unaware of the negative analogies between living cells and machines. But as Victor De Lorenzo, a strong proponent of the engineering perspective and machine metaphors in biology, remarks: "Ultimately, as happens with scientific hypotheses also, all metaphors may be ultimately wrong, but some of them are surely (very) useful." Of course the key question is: useful for what purpose? Science, as we've noted before, has at least two key objectives: One is truthful representation of reality and the other is successful intervention to control and improve reality for human purposes. While the technomorphic metaphors discussed in this chapter may be inadequate for the first task, they appear at the moment to have significant promise for the second, even if they may ultimately need at some point to be revised or rejected in favour of others.

Whether forms of speech such as genome editing, cellular reprogramming, and rewiring will ever come to be regarded as literal or examples of dead metaphor remains uncertain. But it is clear that this language is changing more than just how we talk and think about cells and living organisms, it is impacting how cells and perhaps eventually our own bodies function and behave. Whether it is correct or not to describe these experimental treatments as editing, reprogramming or rewiring, our cells and genomes (and so ultimately ourselves at some more immediate and personal level) are quite likely on the road to being manipulated and altered.

Whatever happens, it should be clear that scientific metaphors are always more than just "lenses" that help us "see" something in an interesting way, or imaginative heuristic tricks for engaging in analogical reasoning. They are also performative and prescriptive: When we describe our cells and our genomes as machines, computers, or programs, for instance, we are implicitly endorsing the idea that it is only right and natural that they be reengineered, redesigned, reprogrammed, and rewired. After all, that's what we do with computers and other bits of engineering. We are also promoting the belief that we actually have the ability to engineer cells and to edit genomes with the same level of precision, reliability, and safety as when we engineer machines and edit texts or computer programs.

For this reason, we need to evaluate these metaphors not only on their contribution to our ability to understand or to manipulate objective reality, but also for their contributions to the construction of the social reality in which we all live. If we continue to describe and think about our *cells* as bits of computer-coded machinery, will we improve the conditions under which we live or will we engineer our *selves* to better conform to the metaphors that currently dominate our imaginations? Will these metaphors further encourage a narrow focus on disease and disability at the cellular and genetic level to the exclusion of attention to the broader social and environmental factors, like pollution and unhealthy lifestyles, which may be more significant causal contributors to non-heritable diseases like cancer?

The use of technomorphic computer and machine metaphors in biology has been a frequent target of criticism. But before closing this chapter I wish to say a few words about the use of agent metaphors.

How the Portrayal of Molecules as Agents Misleads Public Understanding of Cell and Molecular Biology

As noted in earlier chapters, agent metaphors ascribe conscious intent or purposiveness to molecules that are neither alive nor conscious. For instance, proteins and peptides are described as messengers and are said to recruit one another to work in collaborative teams to complete various tasks. As explained in Chapters 3 and 4, these verbal descriptions can assist our recognition and understanding of what molecules, proteins, and cells do. In their essay "Cognition all the way down," the philosopher Daniel Dennett and the biologist Michael Levin argue for the importance of thinking about cells, tissues, and non-human organisms as (unthinking) agents with agendas. This predictably raises a host of questions about whether it is informative or merely misleading to anthropomorphize non-human entities. (The authors' point, by the way, is not so much to anthropomorphize non-human agents, as to demystify our own cognitive abilities as lying on a continuum of quite natural goal-directed behaviours extending down into smaller and unconscious scales.)

But molecules and proteins can also be *visually* represented in ways that more subtly analogize them to human agents, and in a way that is potentially even more misleading for the public. Animated video representations using computer-generated images are an extremely powerful way of explaining how molecular systems like gene expression, protein synthesis, signalling pathways, or CRISPR work. However, they create the impression (either intentionally or not) that these molecules and systems behave with an apparent intelligence because of the specificity with which they are portrayed to interact with their targets. For instance, transcription factors and the CRISPR/Cas system are typically portrayed as proceeding with great purpose to their intended targets in order to carry out their intended function, leading the viewer to ponder how these molecules (or macromolecular complexes in the case of CRISPR/Cas9) know to go straight to the nucleus and start searching for that specific nucleotide sequence.

The answer of course is that they don't know. These animations are misleading because they greatly simplify the vast number of molecules within the cell by focusing on one or two relevant molecules and showing them acting in very direct ways to illustrate the function of interest for which the animation has been created. These are highly selective narratives, and just as an author, historian, or a prosecutor in a court of law describes only the specific behaviour of relevant characters in telling a story, and ignores all the billions of other people on the planet who are also busy doing things and who have internal dialogues of their own, the cell is a busy place packed with millions of molecules that interact or pass one another by, often with multiple copies of the particular molecule of interest present, being bumped constantly about by random Brownian motion; and so, for similar purposes of economy and clarity, the creators of these animations strip away the complexity and focus on just one or two molecules of interest and show them proceeding on the most straightforward path to illustrate the function or phenomenon in question.

But this may encourage the impression that the cell and its components behave like intelligent agents, when in fact the molecules are randomly bumping and jostling one another like people in a crowded stampede. When two or more of them "recognize" one another they may interact or modify one another ("exchange information") in ways that will have consequences for their chance interaction with other molecules further down a long series of largely contingent interactions. (I say *largely contingent* because the cell – especially the eukaryotic cell – is not just a "bag of chemicals" wherein molecules bounce randomly about, but is organized by internal membranes, organelles, and an internal cytoskeleton that increases the chances that molecules required to carry out certain chemical reactions will interact with one another.) The molecular actors themselves are merely obeying the laws of chemistry or following the chemical script in an organized cell environment that has been assembled by natural selection over millions of years, and we need be no more amazed by their apparent purposiveness of activity than we are by the behaviour of leaves when they fall from the tree and pursue with apparent single-mindedness their "goal" of reaching the ground.

To those inclined to ask, "But how do you propose to account for the existence of these *laws of nature* or the *chemical script*?" I remind you that *these are metaphors* and we need not take them literally. Natural selection provides a well-confirmed and effective explanation for how such seemingly purposive and intelligent (teleonomic) molecular systems have come to exist without invoking an *obscurum per obscurius* explanation (explaining the mysterious by appeal to something even more mysterious) involving some supernatural intelligence that designed and created them out of pure nothingness. While belief in such a being may serve pragmatic, emotional, psychological, social, and religious functions, as a scientific hypothesis it explains nothing and provides no helpful guidance to scientists of how they should proceed to learn more about normal and abnormal cell function.

Concluding Remarks

What Is the Significance of Science's Reliance on Metaphor?

What is the significance of science's reliance on metaphor? Does the fact that much of the language of science is non-literal undermine its status as objective knowledge of reality or its ability to help us solve practical problems concerning the world and our health? What should readers keep in mind when they hear or read scientists employing metaphorical language?

Scientific language, especially that which is metaphorical, should be regarded as similar to provisional hypotheses that may require revision or ultimate rejection depending upon what the evidence suggests. We should also be aware that the metaphors scientists use may have quite positive effects for them in their original narrow application, allowing them to think about, understand, and possibly to manipulate some very specific and limited aspect of the world, but that the metaphor may be less adequate when applied to the broader system as a whole.

The aims of science are plural. Some projects seek disinterested knowledge for knowledge's sake (e.g., the origins and history of life), some are more concerned with achieving results by whatever means possible (preventing or curing disease). Metaphors are important tools in both sorts of project, in the first acting like *lenses* that allow us to see or recognize some interesting feature, and in the second like *scalpels* or *tweezers* allowing us to dissect and rearrange life's components to repair them when they are damaged or malfunctioning, or to assemble new kinds of living systems with novel properties.

At best, any metaphor offers only a *partial* and *selective* perspective on reality. We may need, therefore, to adopt several different metaphors if we desire a more complete and objective understanding of things, just as when we attempt to view an object or an issue of debate *metaphorically* "from all sides." And this is especially true when a metaphor that does some useful work within a specific scientific context is transferred into broader discussions of a social, ethical, or political nature, where a different sort of linguistic tool may be more appropriate.

Should Scientists Avoid Using Metaphors?

> Books of life, junk DNA, DNA barcodes: all these images can and have distorted the picture, not least because scientists themselves sometimes forget that they are metaphors. And when the science moves on – when we discover that the genome is nothing like a book, the metaphors tend, nonetheless, to stick. The more vivid the image, the more dangerously seductive and resistant to change it is.

So cautioned Philip Ball, the prolific science writer and former editor for the science journal *Nature*, in 2011. "At the very least," he continues, "metaphor should be admitted into science only after strict examination." As reasonable as this caution may sound, it would likely constitute a prohibition on ever using metaphor – for it puts the cart before the horse, by assuming the metaphors play only an inessential communicative function and are not crucial to the articulation of the very scientific ideas which they help make possible. We can't determine how adequate a metaphor is, except by putting it into use and following the neutral analogies it suggests to see how many turn out to be positive and how many negative. In that way, too, metaphors are like hypotheses; and we can't very well say that scientists should eschew hypotheses. What should be said is that we need to be ready and willing to drop a metaphor when it proves to be inadequate and misleading, or when its deficiencies outweigh its merits. Yet history shows that this is often difficult to do. In other cases, it may not even be necessary, if the metaphor dies and all the negative connotations with it, leaving only the positive analogies attached to the term, at which point it has acquired a new and restricted polysemic meaning. This we saw with the history of the metaphorical term "cell" in Chapter 5.

More recently, Ball seems resigned to the idea that science will always rely to some extent on metaphorical language to help make sense of the world. In his recent book *How to Grow a Human*, Ball pays particular attention to the metaphors central to cell and developmental biology and to how the cultural narratives associated with them influence our scientific understanding of the world. In anticipation of the sort of reader who simply wants the science without the metaphorical framework, he says, "I cannot give you 'just the science', because it already comes with a story attached."

Richard Lewontin advised that "As Arturo Rosenblueth and Norbert Wiener once noted, 'The price of metaphor is eternal vigilance.'" Alternatively, Brendon Larson has recommended that "Diverse metaphors ... act as a prophylactic against reification." By employing multiple metaphors, he suggests, we may be less likely to fall under the thrall of one alone. We have seen that scientists do tend to use multiple metaphors to explore and understand the various different aspects of the systems they study. Just as there is no one uniquely correct universal tool for dealing with the world (hammers are good for some tasks, pliers for others), so there may be no one uniquely correct account (metaphorical or literal) for describing and dealing with the vast complexity of reality.

Some readers might persist in asking whether we can't do better than settling for instrumental utility and insist on a literally, objectively true account of genetics or biology, etc. But what would an "objectively" true account look like? Where would we find this "objective" language and vocabulary? Will nature provide us with it? Will God? And would it even be preferable, were it possible? For what and whose purposes? As others have noted, the object of science is a lot like map-making: We want to accurately depict the terrain in question, but for a multitude of very human purposes, and so a multitude of different sorts of maps of the same terrain are desirable. One for driving from point A to B, another for geology, another for meteorology, another to indicate political ridings or household incomes, etc. Just as there is no one uniquely correct map, it may make sense to forgo the idea that scientists should be aiming at one uniquely correct account of the world. But it does not thereby follow that any metaphor (or map) is just as good as any other. A metaphor must be adequate to the task at hand, and therefore some will be better, and indeed more faithful in their representation and explanation of the facts, than others. So it's worth repeating that we should consider metaphors like hypotheses: They are provisional, partial, relatively useful within specific research contexts, and subject to revision, replacement, or rejection.

Scientific metaphors are also used, and have implications, outside the laboratory. As Brigitte Nerlich, who has devoted her career to understanding how metaphors influence science communication, writes: "public understanding of science is, at least in part, a struggle over metaphors." And if the metaphors are not well suited to the objective of guiding public policy, then our decisions will be negatively impacted. So when the issue of some bit of metaphorical language being used by scientists comes up, it is too simplistic to respond, "It's *only* a metaphor," because metaphors exert significant influence in both scientific communities and society at large. By understanding how the metaphors we use to speak and think about these issues help to set research agenda, frame questions, and define our collective future, scientists and laypeople alike will be better able to make conscious, deliberative, and responsible decisions, rather than being driven forward by the momentum of a set of currently popular metaphors, and while being partially blinded by them.

However, as we become more cognizant of how scientific metaphors can mislead us, we will face another pair of challenges. One is to resist a reactionary embrace of alternative "holistic" metaphors that might sound more humane and spiritually uplifting than the mechanistic and engineering metaphors currently in vogue in many areas of the life sciences, but that promote an unhelpful obscurantism. By now it should be abundantly clear that anti-science, anti-expertise, populist sentiment is especially dangerous in a time of global pandemic, when science's ability to understand the threat and to create safe, effective vaccines is so important. The other challenge is to resist an unrealistic and counterproductive attempt to expunge metaphor from scientific discourse altogether. Rather than reactionary or unrealistic, let us be vigilant.

Summary of Common Misunderstandings

Metaphors are used in science only for communicating complex ideas to non-experts. In fact, scientists use metaphors for at least four different functions. (1) A communicative function, allowing them to talk about novel subject matters with one another and with non-scientists, or to talk of old subject matter in new and interesting ways. This includes rhetorical communication aimed at educating students and persuading other scientists of the correctness and promise of their ideas. (2) A heuristic function, helping scientists to create new concepts, hypotheses, models, and theories about the world that can guide future research. (3) A cognitive function, whereby metaphors enable analogical reasoning that can result in explanations and understanding of how living systems work. And (4), metaphors function almost like tools, helping scientists to mold and manipulate the reality of things they study, as they seek to make the object of the metaphor's target domain more like the object of the source domain. In this sense, metaphors do more than simply change the way we "see" or think about things, they can lead to real material change in the object itself.

It is possible to describe nature in its own objective terms. Nature does not come already labelled with its own objective language. Consequently, scientists must imaginatively *create* terms and descriptions for specific purposes and try them out experimentally, just as they do with hypotheses to see whether or not they will prove to be adequate for the tasks at hand. Many scientific terms and concepts begin as metaphors. Over time, however, these metaphorical terms may be clarified, restricted in meaning, and regarded as literally appropriate, or as dead metaphors.

Scientific terminology perfectly reflects an already articulated reality. Because there are no ways to describe nature in its own objective terms, it may be better to think of scientific terminology and language as instruments rather than pictures that perfectly reflect an already articulated reality. In science, as in the rest of life, some terms are used literally and others metaphorically. Scientists employ language metaphorically for a variety of tasks, as explained above. But it must always be borne in mind that any metaphor is at best a partial and selective perspective on some aspect of reality.

Scientific concepts have clear definitions that are easily understood. This is not always the case. It is important to recognize when a scientific concept is a metaphor or began as a metaphor so as not to be misled by its unintended or inapplicable implications. For example, machine and engineering metaphors in the life sciences encourage teleological thinking that can appear supportive of intelligent design and other creationist accounts of life. The use of social and agential metaphors also dabbles in teleology and anthropomorphism, and can create obstacles to a naturalistic understanding of living systems and organisms. And in some cases a metaphor is useful precisely because it is vague enough to allow specialists from different scientific disciplines to communicate with one another about some general topic while interpreting the details in their own specific ways.

If science relies on metaphors, then scientific knowledge consists of a merely subjective representation of reality. While it is important to recognize that much scientific language is neither literal nor strictly objective in the sense of being nature's own terminology, it does not follow that science is a purely subjective social construction incapable of referring to and offering insight into an objective reality that exists independently of humans. It can achieve this even if, perhaps only if, it is articulated in distinctly human terms that make sense to the scientists who are doing the science and the people it is intended to enlighten. A perfectly objective account of the world expressed in a language unintelligible to humans would be of no scientific value. Science is like a map: It refers to a real world, and a good map accurately tracks the contours of the terrain in question. But to be of any use, it has to represent the world in ways that make sense to the people using it. Metaphors help not only by providing scientists with terms with which to refer to the objects and systems they study, but by offering suggestions as to how these objects and processes might work in familiar terms.

References

Preface

Ball, P. (2019). *How to Grow a Human: Adventures in Who We Are and How We Are Made*. London: William Collins.

Larson, B. M. H. (2011). *Metaphors for Sustainability: Redefining Our Relationship with Nature*. New Haven, CT: Yale University Press.

Chapter 1

Quotation from Bacon: Judson, H. F. (2001). Talking about the genome. *Nature* 409: 769.

Quotation from Parker: Daisha, D. (2001). The tracks of thought. *Nature* 414: 153.

Quotation from Arbib and Hesse: Arbib, M. A. and Hesse, M. B. (1986). *The Construction of Reality*, Cambridge: Cambridge University Press.

Leonard Cohen poem: Cohen, L. (1961). A kite is a victim in *The Spicebox of Life*. Toronto: McLellan and Stewart. Cohen song: Cohen, L. (1969). Bird on the wire, from the album *Songs from a Room*. Columbia Records.

On metaphor generally: Ortony, A. (ed.) (1993 [1979]). *Metaphor and Thought*, 2nd ed. Cambridge: Cambridge University Press.

Lakoff, G. and Johnson, M. (2003 [1980]). *Metaphors We Live By*, 2nd impression with a new afterword. Chicago, IL: University of Chicago Press.

Gibbs, R. W. Jr. (ed.) (2008). *The Cambridge Handbook of Metaphor and Thought*. Cambridge: Cambridge University Press.

On metaphor in science: Black, M. (1962). *Models and Metaphors: Studies in Language and Philosophy*. Ithaca, NY: Cornell University Press.

Black, M. (1979). More about metaphor. In Ortony, A. (ed.), *Metaphor and Thought*, Cambridge: Cambridge University Press, pp. 19–41.

Bradie, M. (1999). Science and metaphor. *Biology and Philosophy* 14: 159–166.

Brown, T. L. (2003). *Making Truth: Metaphor in Science*. Chicago, IL: University of Illinois Press.

Hesse, M. B. (1966). *Models and Analogies in Science*. Notre Dame, IN: University of Notre Dame Press.

Kampourakis, K. (2020). Why does it matter that many biology concepts are metaphors. In Kampourakis, K. and T. Uller (eds.), *Philosophy of Science for Biologists*. Cambridge: Cambridge University Press.

Reynolds, A. S. (2018). *The Third Lens: Metaphor and the Creation of Modern Cell Biology*. Chicago, IL: University of Chicago Press.

See also the collection of papers guest-edited by Vicedo, M. and Walsh, D. (2020). Making sense of metaphor: Evelyn Fox Keller and commentators on language and science, a special issue of *Interdisciplinary Science Reviews* 45(3).

Quotation from Black: "Every metaphor is the tip of a submerged model": Black, M. (1979). More about metaphor. In Ortony, A. (ed.), *Metaphor and Thought*, Cambridge: Cambridge University Press, p. 30.

Metaphors as theory-constitutive: Boyd, R. (1993). Metaphor and theory change: what is "metaphor" a metaphor for? In Ortony, A. (ed.), *Metaphor and Thought*, 2nd ed. Cambridge: Cambridge University Press, pp. 481–532.

Kuhn on metaphors as initiating paradigm shifts: Kuhn, T. S. (1993). Metaphor in science. In Ortony, A. (ed.), *Metaphor and Thought*, 2nd ed. Cambridge: Cambridge University Press, pp. 533–542.

Quotation: "The hope for any metaphor in science . . .": Avise, J. (2001). Evolving genomic metaphors: a new way to look at the language of DNA. *Science* 294 (5540): 87.

Quotation: "The problem ... is not so much that a metaphor is wrong ...": Kueffer, C. and Larson, B. M. H. (2014). Responsible use of language in scientific writing and science communication. *BioScience* 64: 720.

Lewontin on the social and political dimensions of scientific metaphor: Lewontin, R. (1991). *Biology as Ideology: The Doctrine of DNA*. Toronto: House of Anansi Press.

On metaphor choice framing issues: Thibodeau, P. H. and Boroditsky, L. (2011). Metaphors we think with: the role of metaphor in reasoning. *PLoS ONE* 6(2): e16782. Thibodeau, P. H. and Boroditsky, L. (2013). Natural language metaphors covertly influence reasoning. *PLoS ONE* 8(1): e52961.

Quotation: "the metaphors we rely upon may uphold and reinforce outdated scientific paradigms ...": Taylor, C. and Dewsbury, B. M. (2018). On the problem and promise of metaphor use in science and science communication. *Journal of Microbiology and Biology Education* 19(1): 2.

Further evidence of scientists' recognition of metaphor's importance: Olson, M. E., Arroyo-Santos, A., and Vergara-Silva, F. (2019). A user's guide to metaphors in ecology and evolution. *Trends in Ecology & Evolution* 34(7): 605–615.

"Metaphors can be used to highlight and hide or foreground ...": Nerlich, B., Elliott, R., and Larson, B. M. H. (eds.) (2016). *Communicating Biological Sciences: Ethical and Metaphorical Dimensions*. New York: Routledge, pp. 15–16.

Nelkin on promotional metaphors: Nelkin, D. (1994). Promotional metaphors and their popular appeal. *Public Understanding of Science* 3(1): 25–31.

For more on literality, objectivity, and the implications of science's reliance on metaphor for the question of scientific realism, see my previous book, Reynolds, A. (2018). *The Third Lens: Metaphor and the Creation of Modern Cell Biology*. Chicago, IL: University of Chicago Press.

Chapter 2

On the notion of background metaphors: Blumenberg, H. (2010). *Paradigms for a Metaphorology*, translated by R. Savage. Ithaca, NY: Cornell University Press.

A related idea is that of root metaphors: Pepper, S. (1942). *World Hypotheses*. Berkeley, CA: University of California Press.

On the intentional stance: Dennett, D. C. (1989). *The Intentional Stance*. New York: MIT Press.

On the rise of machine metaphors during the scientific revolution: Henry, J. (2008). *The Scientific Revolution and the Origins of Modern Science*, 3rd ed. New York: Palgrave Macmillan.

Westfall, R. (1971). *The Construction of Modern Science: Mechanisms and Mechanics*. Cambridge: Cambridge University Press.

Merchant, C. (1980). *The Death of Nature: Women, Ecology and the Scientific Revolution*. New York: Harper-Collins.

For the application of technological metaphors and analogies to the brain and nervous system: Cobb, M. (2020). *The Idea of the Brain: The Past and Future of Neuroscience*. New York: Basic Books.

On mechanistic explanation in science: Bechtel, W. and Richardson, R. C. (1993). *Discovering Complexity: Decomposition and Localization as Strategies in Scientific Research*. Cambridge, MA: MIT Press.

Craver, C. F. and Darden, L. (2013). *In Search of Mechanisms: Discoveries Across the Life Sciences*. Chicago, IL: University of Chicago Press.

Glennan, S. (2017). *The New Mechanical Philosophy*. Oxford: Oxford University Press.

Quotation from Lewontin, "It seems impossible to do science without metaphors": Lewontin, R. (2001). In the beginning was the word. *Science* 291(5507): 1263–1264, 1263.

Characterization of mechanisms as "entities and activities organized such that they are productive of regular changes from start or set-up to finish or termination conditions": Craver, C. F. and Darden, L. (2013). *In Search of Mechanisms: Discoveries Across the Life Sciences*. Chicago, IL: University of Chicago Press, p. 15.

Ruse on historical evolution of the mechanism concept: Ruse, M. (2005). Darwinism and mechanism: metaphor in science. *Studies in History and Philosophy of Biological and Biomedical Sciences* 36: 285–302.

Quote from Wiener "If the seventeenth and early eighteenth centuries are the age of clocks . . .": Wiener, N. (1961). *Cybernetics or Control and Communication in the Animal and the Machine*, 2nd ed. Cambridge, MA: MIT Press, p. 31.

Quotation from Watson and Crick on genetic code: Watson, J. D. and Crick, F. (1953). Genetical implications of the structure of deoxyribonucleic acid. *Nature* 171: 964–967, 967.

Gamow on machine, language, and code metaphors and DNA: Gamow, G. (1954). Possible relation between deoxyribonucleic acid and protein structures. *Nature* 173: 318. Gamow, G. (1955). Information transfer in the living cell. *Scientific American* 193(4): 70–79.

On the history of computer and code metaphors in genetics: Keller, E. F. (1995). *Refiguring Life: Metaphors of Twentieth-Century Biology*. New York: Columbia University Press. Keller, E. F. (2000). *The Century of the Gene*. Cambridge, MA: Harvard University Press. Keller, E. F. (2002). *Making Sense of Life: Explaining Biological Development with Models, Metaphors, and Machines*. Cambridge, MA: Harvard University Press. Kay, L. (2000). *Who Wrote the Book of Life? A History of the Genetic Code*. Stanford, CA: Stanford University Press. Cobb, M. (2013). 1953: when genes became "information." *Cell* 153: 503–506.

For a critical take on information talk in biology: Griffiths, P. E. (2001). Genetic information: a metaphor in search of a theory. *Philosophy of Science* 68(3): 394–412.

Quotation from Kay, "These particular representations are historically specific and culturally contingent": Kay, L. (2000). *Who Wrote the Book of Life? A History of the Genetic Code*. Stanford, CA: Stanford University Press, p. 2.

On calls for new metaphors to describe genes and genetics: Syzmanski, E. and Scher, E. (2019). Models for DNA design tools: the trouble with metaphors is that they don't go away. *ACS Synthetic Biology* 8: 2635–2641.

For philosophical treatment of scientific understanding and explanation: De Regt, H. (2017). *Understanding Scientific Understanding*. Oxford: Oxford University Press. Woodward, J. (2019). Scientific explanation In Zalta, E. N. (ed.), *Stanford Encyclopedia of Philosophy* (winter 2019 ed.): https://plato.stanford.edu/archives/win2019/entries/scientific-explanation.

On the analogy between maps and theories: Winther, E. (2020). *When Maps Become the World*. Chicago, IL: University of Chicago Press.

On cognitive psychology studies of attribution of agency and purpose to non-human and inanimate objects and processes: Varella, M. A. C. (2018). The biology and evolution of the three psychological tendencies to anthropomorphize biology and evolution. *Frontiers in Psychology* 9: 1839.

Mayr on teleonomy versus teleology: Mayr, E. (1961). Cause and effect in biology. *Science* 134(3489): 1501–1506.

On the stream conception of life: Nicholson, D. (2018). Reconceptualizing the organism: from complex machine to flowing stream. In Nicholson, D. and Dupré, J. (eds.), *Everything Flows: Toward a Processual Philosophy of Biology*. Oxford: Oxford University Press.

On landscape metaphor applied to chromatin: Cremer, T., Cremer, M., Hübner, B., et al. (2015). The 4D nucleome: evidence for a dynamic nuclear landscape based on co-aligned active and inactive nuclear compartments. *FEBS Letters* 589(20): 2931–2943.

Chapter 3

For a nuanced interpretation of what Mendel himself believed his experiments showed: Müller-Wille, S. and Hall, K. (2016). Legumes and linguistics: translating Mendel for the twenty-first century: www.bshs.org.uk/bshs-translations/mendel (translation of and commentary on Mendel's *Experiments on Plant Hybrids* (1866)).

Kampourakis, K. (2020). *Understanding Genes*. Cambridge: Cambridge University Press, especially pp. 27–32.

Quotation from Johannsen "free from any hypotheses …" quoted in Keller, E. F. (2000). *The Century of the Gene*. Cambridge, MA: Harvard University Press, p. 2.

Keller on the discourse of gene agency: Keller, E. F. (1995). *Refiguring Life: Metaphors of Twentieth-Century Biology*. New York: Columbia University Press, p. 8.

Watson and Crick on DNA as code: Watson, J. D. and Crick, F. (1953). Genetical implications of the structure of deoxyribonucleic acid. *Nature*, 171: 964–967, 965.

For an earlier suggestion that the DNA might serve as a kind of "code script," see Schrödinger, E. (2012 [1944]). *What Is Life? With Mind and Matter and Autobiographical Sketches*. Cambridge: Cambridge University Press.

On the central dogma of biology: Crick, F. (1958). On protein synthesis. *Symposia of the Society for Experimental Biology* 12: 138–163.

Quotation from Jacob and Monod, "According to the strictly structural concept . . .": Jacob, F. and Monod, J. (1961). Genetic regulatory mechanisms in the synthesis of proteins. *Journal of Molecular Biology* 3: 318–356, 354.

On the genetic switch metaphor: Jacob, F. (1979). The switch. In *Origins of Molecular Biology: A Tribute to Jacques Monod*, edited by Lwoff, A. and Ullmann, A. New York: Academic Press, pp. 95–108.

On the history of the genetic program metaphor: Peluffo, A. E. (2015). The "genetic program": behind the genesis of an influential metaphor. *Genetics* 200: 685–696.

On microscopic cybernetics: Monod, J. (1972). *Chance and Necessity*. Glasgow: Collins and Sons.

On the concept of program making an honest woman of teleology: Jacob, F. (1973). *The Logic of Life: A History of Heredity*, translated by B. E. Spillman. New York: Pantheon Books, p. 17.

Haldane's remark that teleology is the biologist's mistress quoted in Keller, E. F. (2002). *Making Sense of Life: Explaining Biological Development with Models, Metaphors, and Machines*. Cambridge, MA: Harvard University Press, p. 141.

Keller on the lack of clarity of what the genetic program refers to: Keller, E. F. (2002). *Making Sense of Life: Explaining Biological Development with Models, Metaphors, and Machines*. Cambridge, MA: Harvard University Press.

On genes as convenient "handles": Keller, E. F. (2000). *The Century of the Gene*. Cambridge, MA: Harvard University Press, pp. 141–142.

On the genome as a musical score or recipe: Noble, D. (2006). *The Music of Life: Biology Beyond the Genome*. Oxford: Oxford University Press.

Pistorio, S. (2020). DNA is not a blueprint. *Scientific American*: https://blogs.scientificamerican.com/observations/dna-is-not-a-blueprint.

Lewontin quotation " . . . when biologists speak of genes as 'computer programs,'": Lewontin, R. C. (2000). Foreword. In Oyama, S. (ed.), *The Ontogeny of*

Information: Developmental Systems and Evolution, 2nd ed. Durham, NC: Duke University Press, p. viii.

On genetic essentialism and metaphors: Kampourakis, K. (2020). *Understanding Genes*. Cambridge: Cambridge University Press. Lewontin, R. C. (1991). *Biology as Ideology: The Doctrine of DNA*. Toronto: House of Anansi Press; Lewontin, R. C. (2000). *The Triple Helix: Gene, Organism, and Environment*. Cambridge, MA: Harvard University Press. Hubbard, R. and Wald, E. (1997). *Exploding the Gene Myth: How Genetic Information Is Produced, and Manipulated by Scientists, Physicians, Employers, Insurance Companies, Educators, and Law Enforcers*. Boston, MA: Beacon Press. Moss, L. (2003). *What Genes Can't Do*. Cambridge, MA: MIT Press. Turney, J. (2009). Genes, genomes and what to make of them. In Nerlich, B., Elliott, R., and Larson, B. M. H. (eds.), *Communicating Biological Sciences: Ethical and Metaphorical Dimensions*. London: Routledge.

Quotation from Kay "[metaphors] like the information and code metaphors, are exceptionally potent": Kay, L. (2000). *Who Wrote the Book of Life? A History of the Genetic Code*. Stanford, CA: Stanford University Press, p. 3.

Quotation: "'The Book of Life,' as some have termed the human genome, is actually three books": National Human Genome Research Institute (2003). Budget Justification Submitted to the US Congress, p. 810: https://bit.ly/3i7AYPj.

Quotation from Collins, "I think it is a fairly decent analogy . . . ": Collins, F. (2000). Interview (June 16, 2000) with Bob Abernethy for *Religion & Ethics Newsweekly*: https://to.pbs.org/2S4vkCH.

For more on genome metaphors: Nerlich, B., Dingwell, R., and Clarke, D. D. (2002) The book of life: how the completion of the Human Genome Project was revealed to the public. *Health: An Interdisciplinary Journal for the Social Study of Health, Illness and Medicine* 6(4): 445–469. Pigliucci, M. (2010). Genotype–phenotype mapping and the end of the "genes as blueprint" metaphor. *Philosophical Transactions of the Royal Society B* 365: 557–566. Perrault, S. T. and O'Keefe, M. (2019). New metaphors for new understandings of genomes. *Perspectives in Biology and Medicine* 62(1): 1–19.

Chapter 4

On the application of machine metaphors to cells and proteins: Brown, T. L. (2003). *Making Truth: Metaphor in Science*. Chicago, IL: University of Illinois Press. Reynolds, A. S. (2018). In search of cell architecture: *General Cytology* and early twentieth-century conceptions of cell organization. In Matlin, K. S., Maienschein, J., and Laubichler, M. D. (eds.), *Visions of Cell Biology: Reflections Inspired by Cowdry's General Cytology*. Chicago, IL: University of Chicago Press, pp. 46–72. Reynolds, A. (2018). *The Third Lens: Metaphor and the Creation of Modern Cell Biology*. Chicago, IL: University of Chicago Press. Grote, M. (2019). *Membranes to Molecular Machines: Active Matter and the Remaking of Life*. Chicago, IL: University of Chicago Press.

On the protein machine hypothesis: Kurzyński, M. (1997). Protein machine model of enzymatic reactions gated by enzyme internal dynamics. *Biophysical Chemistry*, 65(1): 1–28, 3.

On the cell as a collection of protein machines: Alberts, B. (1998). The cell as a collection of protein machines: preparing the next generation of molecular biologists. *Cell* 92(3): 291–294, 291. "A true replication machine … ": Alberts, B. (1984). The DNA enzymology of protein machines. *Cold Spring Harbor Symposium on Quantitative Biology* 49: 1–12, 1, 3.

Quotation from Goodsell on differences between molecular and man-made machines: Goodsell, D. S. (2009 [1993]). *The Machinery of Life*, 2nd ed. New York: Springer, p. 9.

Nicholson's "argument from scale": Nicholson, D. J. (2020). On being the right size, revisited: the problem with engineering metaphors in molecular biology. In Holm, S. and Serban, S. (eds.), *Philosophical Perspectives on the Engineering Approach in Biology: Living Machines?* London: Routledge, pp. 40–68, 62.

See also Nicholson's critique of the machine conception of the cell and organism: Nicholson, D. J. (2019). Is the cell *really* a machine? *Journal of Theoretical Biology* 477: 108–126.

Quotation "Some enzyme complexes function literally as machines …": Block, S. M. (1997). Real engines of creation. *Nature*, 386(6622): 217–218.

On "the molecular storm": Hoffmann, P. M. (2012). *Life's Ratchet: How Molecular Machines Extract Order from Chaos*. New York: Basic Books.

Boudry and Pigliucci's critique of engineering metaphors: Boudry, M. and Pigliucci, M. (2013). The mismeasure of machine: synthetic biology and the trouble with engineering metaphors. *Studies in History and Philosophy of Biological and Biomedical Sciences* 44: 660–668.

Johannes Jaeger's YouTube Channel *Beyond Networks: The Evolution of Livings Systems* is a series of videos critical of the machine conception of cells and organisms.

On natural selection as a "blind watchmaker" and a "tinkerer": Dawkins, R. (1986). *The Blind Watchmaker: Why the Evidence of Evolution Reveals a Universe without Design*. London: Norton.

Jacob, F. (1977). Evolution and tinkering. *Science* 196(4295): 1161–1166.

Quotation from Boudry and Pigliucci where "the object of study becomes so remote . . .": Boudry, M. and Pigliucci, M. (2013). The mismeasure of machine: synthetic biology and the trouble with engineering metaphors. *Studies in History and Philosophy of Biological and Biomedical Sciences* 44: 660–668, 667.

On agency metaphors applied to proteins and genomes: Pappas, G. (2005). A new literary metaphor for the genome or proteome. *Biochemistry and Molecular Biology Education* 33(1): 15.

Quotation from Kampourakis "Cell biologists refer to 'signaling' proteins . . . ": Kampourakis, K. (2020). *Understanding Genes*. Cambridge: Cambridge University Press, pp. 111–112.

On "Shakespearean" versus "Newtonian" biology: Lipan, O. and Wong, W. H. (2006). Is the future of biology Shakespearean or Newtonian? *Molecular BioSystems* 2: 411–416.

Chapter 5

For the history of cell theory and metaphor, this chapter draws on my earlier book: Reynolds, A. S. (2018). *The Third Lens: Metaphor and the Creation of Modern Cell Biology*. Chicago, IL: University of Chicago Press.

For additional discussion of metaphors in the history of biochemistry: Reynolds, A. S. (2018). In search of cell architecture: *General Cytology* and early twentieth-century conceptions of cell organization. In Matlin, K. S.,

Maienschein, J., and Laubichler, M. D. (eds.), *Visions of Cell Biology: Reflections Inspired by Cowdry's General Cytology*. Chicago, IL: University of Chicago Press, pp. 46–72.

Quotations from Virchow and Bernard: Virchow, R. (1958). On the mechanistic interpretation of life. In *Disease, Life, and Man: Selected Essays by Rudolf Virchow*, translated and with an introduction by L. J. Rather. Stanford, CA: Stanford University Press, p. 107.

Bernard, C. (1885). *Leçons sur les Phénomènes de la Vie communs aux Animaux et aux Végétaux*, vol. 1. Paris: J-B. Baillière et Fils, p. 358.

For a school exercise comparing the cell to a factory, see AAAS Science Netlink, Comparing a cell to a factory lesson: http://sciencenetlinks.com/student-teacher-sheets/comparing-cell-factory.

The section on metaphor and cell death draws on: Reynolds, A. S. (2014). The deaths of a cell: how language and metaphor influence the science of cell death. *Studies in History and Philosophy of Science Part C: Studies in History and Philosophy of Biological and Biomedical Sciences* 48: 175–184. Quotation from Lockshin "Because computers were just beginning to be talked about . . . ": Maghsoudi, N., Zaketi, Z., and Lockshin, R. A. (2012). Programmed cell death and apoptosis – where it came from and where it is going: from Elie Metchnikoff to the control of caspases. *Experimental Oncology* 34(3): 146–152, 146.

On programmed cell death in bacteria: Nedelcu, A., Driscoll, W., Durand, P., et al. (2011). On the paradigm of altruistic suicide in the microbial world. *Evolution* 65(1): 3–20. Tanouchi, Y., Lee, A. J., Meredith, H., and You, L. (2013). Programmed cell death in bacteria and implications for antibiotic therapy. *Trends in Microbiology* 21(6): 265–270.

On a "cell-wide web of communication": Duan, J., Navarro-Dorado J., Clark, J. H., et al. (2019). The cell-wide web coordinates cellular processes by directing site-specific Ca^{2+} flux across cytoplasmic nanocourses. *Nature Communications* 10: 2299.

On the history of stem cells: Dröscher, A. (2014). Images of cell trees, cell lines, and cell fates: the legacy of Ernst Haeckel and August Weismann in stem cell research. *History and Philosophy of the Life Sciences* 36(2): 157–186.

On metaphors, societal values, and fertilization: Martin, E. (1991). The egg and the sperm: how science has constructed a romance based on stereotypical male–female roles. *Signs* 16(3): 485–501. Gilbert, S. and Pinto-Correia, C. (2017). *Fear, Wonder, and Science in the New Age of Reproductive Biotechnology*. New York: Columbia University Press. Beldecos, A., Bailey, S., Gilbert, S., et al. (1988). The importance of feminist critique for contemporary cell biology. *Hypatia* 3(1): 61–76.

For the attribution of gender to cells: Richardson, S. (2013). *Sex Itself: The Search for Male and Female in the Human Genome*. Chicago, IL: University of Chicago Press.

Just and Lillie's recognition of egg activity in fertilization: Lillie, F. R. and Just, E. E. (1924). Fertilization. In Cowdry, E. V. (ed.), *General Cytology*. Chicago, IL: Chicago University Press, pp. 451–536, 456.

On gendered metaphors and glial cells: Upchurch, M. and Fojotvá, S. (2009). Women in the brain: a history of glial cell metaphors. *NWSA Journal* 21 (2): 1–20.

On racist and sexist assumptions of scientific analogies: Stepan, N. L. (1986). Race and gender: the role of analogy in science. *Isis* 77(2): 261–277.

On bad use of metaphors and use of bad metaphors: Kampourakis, K. (2016). The bad use of metaphors and the use of bad metaphors. *Science & Education* 25: 947–949.

On ingrained analogies and the importance of inclusivity of language and under-represented groups in science: Sullivan-Clarke, A. (2019). Misled by metaphor: the problem of ingrained analogy. *Perspectives on Science* 27(2): 153–170. Kueffer, C. and Larson, B. M. H. (2014). Responsible use of language in scientific writing and science communication. *Bioscience* 64(8): 719–724. Taylor, C. and Dewsbury, B. M. (2018). On the problem and promise of metaphor use in science and science communication. *Journal of Microbiology and Biology Education* 19(1): DOI: https://doi.org/10.1128/jmbe.v19i1.1538.

On the importance of choosing scientific terminology carefully: Elliott, K. C. (2017). *A Tapestry of Values: An Introduction to Values in Science*. Oxford: Oxford University Press, especially chapter 6

Chapter 6

For a very clear and accessible exposition of evolutionary biology, I recommend: Kampourakis, K. (2020). *Understanding Evolution*. Cambridge: Cambridge University Press.

A pioneering analysis of Darwin's use of metaphor is found in Young, R. M. (1985). *Darwin's Metaphor: Nature's Place in Victorian Culture*. Cambridge: Cambridge University Press.

Quotation from Ruse "starting with the key notions of selection and struggle . . .": Ruse, M. and Richards, R. J. (2016). *Debating Darwin*. Chicago, IL: University of Chicago Press, p. 48.

Quotations from Darwin: Darwin, C. (1859). *On the Origin of Species by Means of Natural Selection or The Preservation of Favoured Races in the Struggle for Life*, 1st ed. London: John Murray and 6th ed. (1872), both available at Darwin Online: http://darwin-online.org.uk/EditorialIntroductions/Freeman_OntheOriginofSpecies.html.

For an example of confusion over Darwin's phrase "survival of the fittest," see Cimons, M. (2020). "Friendliest," not fittest, is key to evolutionary survival, scientists argue in book. *Washington Post,* July 20. (Review of Hare, B., and Woods, V. (2020). *Survival of the Friendliest: Understanding Our Origins and Rediscovering Our Common Humanity*. New York: Random House.)

For the history of Woese and the molecular approach to the Tree of Life I rely on Quammen, D. (2018). *The Tangled Tree: A Radical New History of Life*. New York: Simon & Shuster.

Quotation from Doolittle "we must now admit that any tree ... ": Doolittle, W. F. (2000). Uprooting the tree of life. *Scientific American* February: 90–95, 94.

On trees as networks: Morrison, D. A. (2014). Is the tree of life the best metaphor, model, or heuristic for phylogenetics? *Systematic Biology* 63(4): 628–638.

New Scientist (2009). Darwin was wrong: Cutting down the tree of life. Cover of *New Scientist* 2692.

Quotation from Doolittle: "The tree of life is not something that exists in nature . . ." Doolittle, W. F. (1999). Phylogenetic classification and the universal tree. *Science* 284: 2124–2128.

On selfish genes: Dawkins, R. (1989[1976]). *The Selfish Gene*, new ed. Oxford: Oxford University Press.

Quotation from Dawkins "swarm in huge colonies, safe inside gigantic lumbering robots . . . ": Dawkins, R. (1989 [1976]). *The Selfish Gene*, new ed. Oxford: Oxford University Press, pp. 19–20.

On the cooperative gene: Dawkins, R. (2006) *The Selfish Gene*, 30th anniversary ed. Oxford: Oxford University Press, pp. ix–x.

On spandrels: Gould, S. J. and Lewontin, R. C. (1979). The spandrels of San Marco and the panglossian paradigm: a critique of the adaptationist programme. *Proceedings of the Royal Society of London B* 205: 581–598.

Quotation from Ruse "One has to transcend dichotomies of objective/subjective, discovered/created, description of reality/social construction . . .": Ruse, M. (2006). *Darwinism and Its Discontents*. Cambridge: Cambridge University Press, p. 212.

Readers might also find of interest: Radick, G. (2009). Is the theory of natural selection independent of its history? In Hodge, J. and Radick, G. (eds.), *The Cambridge Companion to Darwin*, 2nd ed. Cambridge: Cambridge University Press, pp. 147–172.

For scientific objectivity metaphorically construed as the view from nowhere: Nagel, T. (1986). *The View from Nowhere*. Oxford: Oxford University Press.

Chapter 7

Especially relevant for the themes of this chapter and the last is: Olson, M. E., Arroyo-Santos, A., and Vergara-Silva, F. (2019). A user's guide to metaphors in ecology and evolution. *Trends in Ecology & Evolution* 34(7): 605–615.

For the history of ecology, this chapter draws chiefly on: Kricher, J. (2009). *The Balance of Nature: Ecology's Enduring Myth*. Princeton, NJ: Princeton University Press. Allaby, M. (1996). *Basics of Environmental Science*. New York: Routledge.

For Linnaeus and the economy of nature: Müller-Wille, S. (2003). Nature as a marketplace: Linnaean botany as a political economy. *History of Political Economy* 35(Suppl.1): 154–172.

Quotation from Engels: "The whole Darwinian theory of the struggle for existence is simply the transference from society to animate nature of Hobbes' theory . . .": Radick, G. (2009). Is the theory of natural selection independent of its history? In Hodge, J. and Radick, G. (eds.), *The Cambridge Companion to Darwin*, 2nd ed. Cambridge: Cambridge University Press, pp. 147–172, 158–159.

For Darwin and the economy of nature: Pearce, T. (2010). "A great complication of circumstances": Darwin and the economy of nature. *Journal of the History of Biology* 43(3): 493–528.

For ecology as defined by Haeckel: "the collective science of the relations of organisms to their environment": Haeckel, E. (1866). *Generelle Morphologie*, vol. 2. Berlin: Reimer, p. 286.

On ecological niches: Pocheville, A. (2015). The ecological niche: history and recent controversies. In Heams, T., Huneman, P., Lecointre, P., and Silberstein, M. (eds.), *Handbook of Evolutionary Thinking in the Sciences*, Dordrecht: Springer, pp. 547–586.

Quotations from Lewontin "the use of the metaphor of a niche implies a kind of ecological space with holes in it . . ." and "Just as there can be no organism without an environment . . . ": Lewontin, R. C. (2000). *The Triple Helix: Gene, Organism and Environment*. Cambridge, MA: Harvard University Press, pp. 45, 48.

On niche-construction theory: Odling-Smee, J., Laland, K., and Feldman, M. (2003). *Niche Construction: The Neglected Process in Evolution*. Princeton, NJ: Princeton University Press.

On the ecosystem concept: Tansley, A. G. (1935). The use and abuse of vegetational concepts and terms. *Ecology* 16(3): 284–307.

"to include with the biome all the physical and chemical factors . . . ": Allaby, M. (1996). *Basics of Environmental Science*. New York: Routledge, p. 150.

Quotation from Tansley "In an ecosystem the organisms and the inorganic factors alike are *components* ...": Tansley, A. G. (1935). The use and abuse of vegetational concepts and terms. *Ecology* 16(3): 284–307, 306.

Quotation from Cuddington "nature is orderly and beneficent ...": Cuddington, K. (2001). The "balance of nature" metaphor and equilibrium in population ecology. *Biology & Philosophy* 16: 463–479, 466.

Larson on the "balance of nature": Larson, B. M. H. (2011). *Metaphors for Environmental Sustainability: Redefining Our Relationship with Nature*. New Haven, CT: Yale University Press, p. 6.

Kricher on the "balance of nature": Kricher, J. (2009). *The Balance of Nature: Ecology's Enduring Myth*. Princeton, NJ: Princeton University Press, p. 16.

Quotations from Kricher that the balance of nature "has always been a fuzzy, poorly defined idea ..." and of "little value in evolution and ecology": Kricher, J. (2009). *The Balance of Nature: Ecology's Enduring Myth*. Princeton, NJ: Princeton University Press, pp. 19, 23.

On current tensions about the language of global warming and climate change: Brown, T. L. (2003). *Making Truth: Metaphor in Science*. Chicago, IL: University of Illinois Press, chapter 9.

Quotation from Allaby "To some non-scientists, however, 'ecology' suggests a kind of stability ... ": Allaby, M. (1996). *Basics of Environmental Science*. New York: Routledge, p. 9.

"*Ecological health* is a nebulous concept that should be expunged from the vocabulary ...": Lancaster, J. (2000). The ridiculous notion of assessing ecological health and identifying the useful concepts underneath. *Human and Ecological Risk Assessment: An International Journal* 6(2): 213–222, 214.

On category mistakes: Ryle, G. (2002 [1949]). *The Concept of Mind*. Chicago, IL: University of Chicago Press.

For criticism of "ecological health": Lackey, R. (2001). Values, policy, and economic health. *BioScience* 51(6): 437–443.

Quotations from Hutton "solid body of earth, an aqueous body of sea ...": Hutton, J. (1795). *Theory of the Earth: With Proofs and Illustrations*, Vol. 1. Edinburgh: www.gutenberg.org/files/12861/12861-h/12861-h.htm.

On Hutton's metaphors: Norwick, S. (2002). Metaphors of Nature in James Hutton's "Theory of the Earth with Proofs and Illustrations." *Earth Sciences History* 21(1): 26–45.

On the Gaia hypothesis: Lovelock, J. E. (1979). *Gaia: A New Look at Life on Earth.* Oxford: Oxford University Press.

Quotations from Lovelock, "Organisms and their environment evolve as a single, self-regulating system" and "organisms and their material environment evolve as a single, coupled system": Lovelock, J. E. (2003). The living Earth. *Nature* 426(18): 769–770, 769.

On Daisyworld: Watson, A. J. and Lovelock, J. E. (1983). Biological homeostasis of the global environment: the parable of Daisyworld. *Tellus Series B* 35(4): 286–289.

Lovelock, J. E. and Margulis, L. (1974). Atmospheric homeostasis by and for the biosphere: the Gaia hypothesis. *Tellus Series A* 26(1–2): 2–10.

On Darwinizing Gaia: Doolittle, W. F. (2017). Darwinizing Gaia. *Journal of Theoretical Biology* 434: 11–19.

Doolittle W. F. and Inkpen, S. A. (2018). Processes and patterns of interaction as units of selection: an introduction to ITSNTS thinking. *PNAS* 115(16): 4006–4014.

The Great Law of Peace of the Iroquois/Haudenosaunee Confederacy, Wampum #28: www.ganienkeh.net/thelaw.html.

Chapter 8

For the nature of biomedicine: Valles, S. (2020). Philosophy of biomedicine. In Zalta, E. N. (ed.), *The Stanford Encyclopedia of Philosophy* (summer 2020 ed.): https://plato.stanford.edu/archives/sum2020/entries/biomedicine.

On metaphors of illness, medicine, and pandemic: Sontag, S. (2001). *Illness as Metaphor and AIDS and Its Metaphors*. New York: Farrar, Strauss and Giroux, and Picador. Van Rijn-van Tongeren, G. W. (1997). *Metaphors in Medical Texts*. Amsterdam: Rodopi. Nerlich, B. (2020). Metaphors in times of a global pandemic. In Wuppuluri, S. and Grayling, A. C. (eds.), *Words and Worlds: Use and Abuse of Analogies and Metaphors within Sciences and Humanities*. New York: Springer.

Readers should be aware of Brigitte Nerlich's blog, *Making Science Public*, where she writes regularly about the role of metaphors in communicating genomics, synthetic biology, climate change, and other science to the public: https://blogs.nottingham.ac.uk/makingsciencepublic.

Some of the first attempts to address ethical issues involving genetic engineering of human embryos are: Glover, J. (1984). *What Sort of People Should There Be?* New York: Penguin. Glover, J. (2007). *Choosing Children: The Ethical Dilemmas of Genetic Intervention*. Oxford: Oxford University Press.

For a more recent discussion, see Baylis, F. (2019). *Altered Inheritance: CRISPR and the Ethics of Human Genome Editing*. Cambridge, MA: Harvard University Press.

For the perspective of a scientist involved in the creation of a new powerful genome editing technology, see Doudna, J. and Sternberg, S. H. (2017). *A Crack in Creation: Gene Editing and the Unthinkable Power to Control Evolution*. New York: Houghton Mifflin Harcourt.

For the history of genetic engineering and related developments, I have relied largely on Morange, M. (2020). *The Blackbox of Biology: A History of the Molecular Revolution*, translated by M. Cobb. Cambridge, MA: Harvard University Press.

Stevens, H. (2016). *Biotechnology and Society: An Introduction*. Chicago, IL: University of Chicago Press.

On genome editing and gene targeting: Urnov, F. D., Rebar, E. J., Holmes, M. C., Zhang, H. S., and Gregory, P. D. (2010). Genome editing with engineered zinc finger nucleases. *Nature Reviews Genetics* 11(9): 636–646. See also Bak, R., Gomez-Espina, N., and Porteus, M. (2018). Gene editing on center stage. *Trends in Genetics* 34(8): 600–611.

Quotation from Doudna and Charpentier, "Our study further demonstrates that the Cas9 endonuclease family can be programmed …": Jinek, M., Chylinski, K., Fonfara, I., et al. (2012). A programmable dual-RNA-guided DNA endonuclease in adaptive bacterial immunity. *Science* 337: 816–820, 816.

Quotations from Doudna, J. (2015). How CRISPR lets us edit our DNA. TED Talk: www.youtube.com/watch?v=TdBAHexVYzc.

Doudna quotations, "malleable as a piece of literary prose . . . " in Baylis, F. (2019). *Altered Inheritance: CRISPR and the Ethics of Human Genome Editing*. Cambridge, MA: Harvard University Press, p. 11; original in Doudna, J. and Sternberg, S. H. (2017). *A Crack in Creation: Gene Editing and the Unthinkable Power to Control Evolution*. New York: Houghton Mifflin Harcourt, p. 90.

For criticism of the molecular scissors description of CRISPR: Hortle, E. (2019). Why the "molecular scissors" metaphor for understanding CRISPR is misleading. *The Conversation*: https://theconversation.com/why-the-molecular-scissors-metaphor-for-understanding-crispr-is-misleading-119812.

On the "fruitful friction" of multiple metaphors: Larson, B. M. H. (2009). Optimizing friction between alternative genomic metaphors: how much plurality is enough? *Genomics, Society and Policy* 5(3): 20–28.

Morange on the implications of the genome editing metaphor: Morange, M. (2020). *The Blackbox of Biology: A History of the Molecular Revolution*, translated by M. Cobb. Cambridge, MA: Harvard University Press, p. 308.

On Cas9 as a molecular scalpel: Mummidivarapu, N. (2020). The promise of CRISPR: a rhetorical analysis of dominant metaphors in an emerging technology. Unpublished MPhil. thesis, Department of History and Philosophy of Science, University of Cambridge.

On CRISPR and targeting metaphors: O'Keefe, M., Perrault, S., Halpern, J., et al. (2015). "Editing" genes: a case study about how language matters in bioethics. *The American Journal of Bioethics* 15(12): 3–10. Oftedal, G. (2018). The role of "missile" and "targeting" metaphors in nanomedicine. *Philosophia Scientiae, Travaux d'Histoire des Sciences et Philosophie/Studies in History of Sciences and Philosophy* 23(1): 39–55.

Quotation from Urnov: "If human embryo editing for reproductive purposes or germline editing were space flight": Ledford, H. (2020). CRISPR gene editing in human embryos wreaks chromosomal mayhem. *Nature* 583: 17–18.

For the scientific response to editing of human twin embryos: Lanphier, E., Urnov, F., Haecker, S. E., Werner, M., and Smolenski, J. (2015). Don't edit the human germline. *Nature* 519(7544): 410–411.

On programmable enzymes: Abudayyeh, O. O., Gootenberg, J. S., Franklin, B., et al. (2019). A cytosine deaminase for programmable single-base RNA editing. *Science* 365(6451): 382–386. Reardon, S. (2020). A new twist on gene editing. *Nature* 578: 25–27.

On metaphors in drug design, the following blog post by Derek Lowe (and comments) is very informative: Lowe, D. (2011). Block that metaphor: metaphors, good and bad. *In the Pipeline* blog, March 17: https://blogs.sciencemag.org/pipeline/archives/2011/03/17/block_that_metaphor.

On the renaming of the journal *Cloning* to *Cloning and Stem Cells*: Wilmut, I. and Taylor, J. (2018). Cloning after Dolly: editorial. *Cellular Reprogramming* 20(1): DOI: 10.1089/cell2018.29011.psiw.

On the epigenetic landscape metaphor, see: Baedke, J. (2018). *Above the Gene, Beyond Biology: Toward a Philosophy of Epigenetics*. Pittsburgh, PA: University of Pittsburgh Press.

Squier, S. M. (2017). *Epigenetic Landscapes: Drawings as Metaphor*. Durham, NC: Duke University Press.

Description of reprogramming from www.nature.com/subjects/reprogramming.

Developmental landscape and reprogramming illustration from: Takahashi, K. (2012). Cellular reprogramming: lowering gravity on Waddington's epigenetic landscape. *Journal of Cell Science* 125(11): 2553–2560.

CRISPR used to create "programmable artificial transcription factors": Lo, A. and Qi, L. (2017). Genetic and epigenetic control of gene expression by CRISPR-Cas systems. *F1000Research* 6: DOI: 10.12688/f1000research.11113.1.

For a current and informative example of engineering, reprogramming, and circuit metaphors: Wang, N. B., Beitz, A. M., and Galloway, K. E. (2020). Engineering cell fate: applying synthetic biology to cellular reprogramming. *Current Opinion in Systems Biology* 24: 18–31.

For more on metaphors in immunology: Swiatczak, B. and Tauber, A. I. (2020). Philosophy of Immunology. In Zalta, E. N. (ed.), *The Stanford Encyclopedia of Philosophy* (summer 2020 ed.): https://plato.stanford.edu/archives/sum2020/entries/immunology.

On Microsoft computer scientists collaborating with cancer biologists: Linn, A. (n. d.). How Microsoft computer scientists and researchers are working to "solve" cancer: https://news.microsoft.com/stories/computingcancer.

For the "hallmarks of cancer": Hanrahan, D. and Weinberg, R. (2011). Hallmarks of cancer: the next generation. *Cell* 144(5): 646–674.

On rewiring immune cells to attack cancer: Tometz, K. (2016). Northwestern researchers rewire cells to attack cancer. Wttw Chicago PBS: https://news.wttw.com/2016/12/13/northwestern-researchers-rewire-cells-attack-cancer.

On rewiring/rerouting/hijacking cell signal pathways and GEARS: Krawczyk, K., Scheller, L., Kim, H., and Fussenegger, M. (2020). Rewiring of endogenous signaling pathways to genomic targets for therapeutic cell reprogramming. *Nature communications* 11(1): 1–9.

"all metaphors may be ultimately wrong, but some of them are surely (very) useful": De Lorenzo, V. (2018). Evolutionary tinkering vs. rational engineering in the time of synthetic biology. *Life Sciences, Society and Policy* 13(13): 16: DOI: 10.1186/s40504-018-0086-x. This is an adaptation of a famous expression attributed to the statistician George Box that "All models are wrong, but some are useful." Box, G. E. P. (1976). Science and statistics. *Journal of the American Statistical Association* 71(356): 791–799: DOI: 10.1080/01621459.1976.1048094.

On different aims of science: Hacking, I. (1983). *Representing and Intervening: Introductory Topics in the Philosophy of Natural Science*. Cambridge: Cambridge University Press. Potochnik, A. (2017). *Idealization and the Aims of Science*. Chicago, IL: University of Chicago Press.

Dennett and Levin on treating cells and other things as cognitive agents: Levin, M. and Dennett, D. C. (2020). Cognition all the way down: biology's next great horizon is to understand cells, tissues and organs as agents with agendas (even if unthinking ones). *Aeon*, October 13: https://aeon.co/essays/how-to-understand-cells-tissues-and-organisms-as-agents-with-agendas.

For films depicting subcellular molecules as agents, see: *Journey Inside the Cell,* produced by the Discovery Institute, a creationist think tank that promotes "intelligent design" theory as a religion-friendly alternative to evolutionary accounts of cellular and organismal origins; *Cell Signals,* created by Cold

Spring Harbor Laboratory and Interactive Knowledge, Inc.; and *Genome Editing with CRISP-Cas9*, created by the McGovern Institute. All are easily found on YouTube.

On the impact of editing, engineering, and agency metaphors on medicine and society: Nelson, S. C., Yu, J.-H., and Ceccarelli, L. (2015). How metaphors about the genome constrain CRISPR metaphors: separating the "text" from its "editor." *The American Journal of Bioethics* 15(12): 60–62. Ceccarelli, L. (2018). CRISPR as agent: a metaphor that rhetorically inhibits the prospects for responsible research. *Life Sciences, Society and Policy* 14: 24. McLeod, C. and Nerlich, B. (2017). Synthetic biology, metaphors and responsibility. *Life Sciences, Society and Policy* 13(13); and all the articles in McLeod, C., and Nerlich, B. (eds.) (2018). Synthetic biology: how the use of metaphors impacts on science, policy and responsible research. Special Issue of *Life Sciences, Society and Policy* 14.

On the influence of environmental factors (and their interaction with genetics): Kampourakis, K. (2020). *Understanding Genes*. Cambridge: Cambridge University Press.

Concluding Remarks

Quotations from Ball: Ball, P. (2011). A metaphor too far. *Nature*: DOI: 10.1038/news.2011.115.

Ball, P. (2019). *How to Grow a Human: Adventures in Who We Are and How We Are Made*. London: William Collins, p. xi.

Quotation from Lewontin " … the price of metaphor is eternal vigilance": Lewontin, R. C. (2001). In the beginning was the word. *Science* 291(5507): 1263–1264. (A review of Kay, L. (2000). *Who Wrote the Book of Life?* Stanford, CA: Stanford University Press.)

Quotation from Larson "Diverse metaphors … act as a prophylactic against reification": Larson, B. M. H. (2009). Optimizing friction between alternative genomic metaphors: how much plurality is enough? *Genomics Society and Policy* 5(3): 20–28, 22.

On the similarity between science and map-making: Winther, E. (2020). *When Maps Become the World*. Chicago, IL: University of Chicago Press.

Quotation from Nerlich "public understanding of science is …": Nerlich, B., Dingwell, R., and Clarke, D. D. (2002). The book of life: how the completion of the Human Genome Project was revealed to the public. *Health: An Interdisciplinary Journal for the Social Study of Health, Illness and Medicine* 6 (4): 445–469, 465.

Index

Locators in *italic* refer to figures